Pentacoordinated Phosphorus

Pentacoordinated Phosphorus

Volume II
Reaction Mechanisms

ROBERT R. HOLMES

University of Massachusetts
Amherst, Massachusetts 01003

ACS Monograph **176**

SERIES

AMERICAN CHEMICAL SOCIETY

WASHINGTON, D. C.   1980

CD/3/19

Library of Congress CIP Data

Holmes, Robert Richard, 1928–
  Pentacoordinated phosphorus.
  (ACS monograph: 175–176 ISSN 0065-7719)

  Includes bibliographies and indexes.

  CONTENTS: v. 1. Structure and spectroscopy.—
v. 2. Reaction mechanisms.
  1. Phosphorus compounds. 2. Coordination compounds. I. Title. II. Series: American Chemical Society.
ACS monograph; 175–176.

QD181.P1H64   546'.7122   80–26302
ISBN 0–8412–0458–6 (v. 1) ACMOAG 175 1–479
                                        (1980)
ISBN 0–8412–0528–0 (v. 2) ACMOAG 176 1–237
                                        (1980)

GENERAL INTRODUCTION

American Chemical Society's Series of Chemical Monographs

By arrangement with the interallied Conference of Pure and Applied Chemistry, which met in London and Brussels in July 1919, the American Chemical Society undertook the production and publication of Scientific and Technologic Monographs on chemical subjects. At the same time it was agreed that the National Research Council, in cooperation with the American Chemical Society and the American Physical Society, should undertake the production and publication of Critical Tables of Chemical and Physical Constants. The American Chemical Society and the National Research Council mutually agreed to care for these two fields of chemical progress.

The Council of the American Chemical Society, acting through its Committee on National Policy, appointed editors and associates to select authors of competent authority in their respective fields and to consider critically the manuscripts submitted. Since 1944 the Scientific and Technologic Monographs have been combined in the Series. The first Monograph appeared in 1921, and up to 1972, 168 treatises have enriched the Series.

These Monographs are intended to serve two principal purposes: first to make available to chemists a thorough treatment of a selected area in form usable by persons working in more or less unrelated fields to the end that they may correlate their own work with a larger area of physical science; secondly, to stimulate further research in the specific field treated. To implement this purpose the authors of Monographs give extended references to the literature.

Contents

Preface

I HAVE WRITTEN THIS monograph for two reasons: to fill a need brought about by the rapid development of the structural and dynamic stereochemistry of pentacoordinated phosphorus and, to treat the subject matter from various points of view in an effort to suggest new research approaches which might stimulate the future development of this fascinating topic.

Much of the subject matter covered in this monograph did not exist a decade ago. The recent developments, particularly in the area of theoretical interpretation, have led to the correlation of structural and spectroscopic data. The subject in a sense has a unique quality. As structural principles on newly synthesized pentacoordinated phosphorus compounds were evolving, application of the resulting pentacoordinate models to a wide variety of reaction mechanisms of phosphorus compounds, including biologically important nucleotidyl and phosphoryl transfer enzyme systems, proved highly successful. Thus, a concerted effort arose involving chemists from several major disciplines.

The treatment is presented in two volumes. Volume I deals with the experimental findings leading to the structural principles for pentacoordinated phosphorus compounds; Volume II centers attention on bonding and the application of principles in the consideration of reaction mechanisms of phosphorus compounds proceeding via pentacoordinated states. The latter volume should prove suitable for adoption as a class text for courses covering these topics either at the advanced undergraduate or graduate level.

The two volumes cover six chapters and contain two appendices. Volume I comprises the first four chapters and the appendices. Volume II contains the remaining two chapters. Sufficient background is given in each chapter to give the reader, unfamiliar with a particular area, an appreciation of the analysis of the accompanying data. Thus, the range of applicability of the technique under discussion and the significance of the associated concepts and principles should be readily apparent. Ample cross-referencing to other chapters is given in which the same topic has been treated by another approach. These comparisons provide an assessment of the relative definitiveness of the data in establishing a particular point. In cross-referencing, the chapter or volume number is cited first, if necessary, followed by the page number.

The order of presentation of subject matter is designed to provide the background needed to cope with the rationale used in entertaining specific mechanistic proposals discussed in Volume II. Most of the structural information from experimental approaches are described in Chapters 2 and 3 of Volume I. The former chapter centers on ground-state phosphorane geometries and their structural distortions from idealized symmetries. These are obtained from diffraction techniques and vibrational spectroscopy. Ground-state distortions are placed on a quantitative basis and representative ORTEP drawings are included to provide a visual perspective. A presentation of the structures of phosphoranyl radicals from ESR spectroscopy also is given. Chapter 3 of Volume I focuses attention on fluxional behavior and ligand exchange mechanisms derived from temperature-dependent NMR spectra.

More theoretical aspects are covered in Chapter 4 of Volume I and Chapter 1 of Volume II and provide insight into the origin of stereochemical nonrigidity for pentacoordinated phosphorus molecules. Based on potential functions established from vibrational data, an extensively parameterized molecular mechanics model is presented in Chapter 4 which is capable of closely simulating phosphorane structures. The latter model is applied in Chapter 2 of Volume II to the active site mechanism of RNase action on RNA fragments. Theoretical treatments of bonding in the first chapter of Volume II lead to the development of a model for estimating the relative energy of all trigonal bipyramidal and square bipyramidal isomers for a given phosphorane, which reproduces measured NMR activational energies for intramolecular ligand exchange to within ±1.5 kcal/mol. The usefulness of this model becomes apparent in Chapter 2 of Volume II in considering the relative merits of possible pentacoordinate transition states invoked in phosphorus reaction mechanisms.

The last chapter of Volume II is the most extensive chapter and contains a discussion of pentacoordinated intermediates postulated in phosphoryl and nucleotidyl transfer enzymes. A detailed presentation of the action of the well-characterized RNase enzyme system on RNA is included as is the mechanism of DNA replication and RNA transcription. In addition to the major phosphorus reactions concerned with acyclic and cyclic phosphorane intermediates, reactions of tricoordinated phosphorus leading to pseudo-pentacoordinate intermediates containing an electron pair and reactions of phosphoranyl radicals containing an odd electron are treated.

As with any area of investigation under rapid development, some topics are open-ended and some may be expressed in different forms. In reviewing these areas, I have made a special effort to point them out. In an attempt to reach some conclusion at these junctures, tentative though they may be, I have taken the liberty of imposing my view on the reader. In addition, for the benefit of those carrying out research in IR and Raman spectroscopy, a large number of tables and spectra summarizing vibrational analyses of pentacoordinate molecules is included in Appendix I of Volume I.

I am very grateful to all who have helped me in the preparation of this book. Included are most of the scientists whose research forms the basis of this book and graduate students at the University of Massachusetts who supplied helpful suggestions during a course covering much of the subject matter. The writing began at the National Institutes of Health, Bethesda, Maryland, continued at the Lehrstuhl für Anorganische Chemie, Braunschweig Universität, West Germany, and was completed at the Laboratoire de Chimie Quantique, Université Louis Pasteur, Strasbourg, France. I thank

Dr. Ira W. Levin, Professor Reinhard Schmutzler, and Dr. Alain Veillard for their hospitality and enlightened discussions which aided the development of this work. Special recognition is given to Professor Joan A. Deiters, Vassar College, whose research with me over the past decade formed the basis of many of the more quantitative aspects of the topics covered and to Dr. Roberta O. Day who carried out many of the x-ray structural studies and prepared most of the drawings and graphs. I particularly wish to thank Patricia Barschenski for typing the manuscript and my wife, Joan, for her help in proofreading and constructing computer drawings as well as for her patience and encouragement during the writing of this book. I enjoyed working with the editorial staff of the American Chemical Society—Candace Deren, Joan Comstock, and Gene Thornton. Their expert help made the actual production of the monograph a pleasant task.

University of Massachusetts   ROBERT R. HOLMES
Amherst, Massachusetts 01003

August, 1980

Chapter

1

Electronic Structure and Polytopal Rearrangement

IN OUTLINING THE STRUCTURAL and spectroscopic aspects of pentacoordinate phosphorus and accompanying principles in the preceding volume, frequent reference to theoretical justification was made to gain a deeper appreciation for their underlying significance. In this chapter, we survey the more important theoretical advances. It will be seen that both the gross bonding features and the presence of facile intramolecular ligand exchange processes, or polytopal rearrangements as they frequently are called, can be interpreted satisfactorily.

Theoretical treatments are in accord on the following points: that the trigonal bipyramid is more stable than the square pyramid and that the axial bonds in the trigonal bipyramid are predicted to be longer than equatorial linkages containing like ligands. However, these studies deal exclusively with simple substituents (monodentate). In fact, in the more rigorous treatments, the calculations are confined to PF_5, the nonexistent PH_5 molecule, and derivatives between these two extremes.

In addition to these areas of general agreement, those studies that center on a description of electronic structure ascribe an electron-deficient character to axial bonds in the trigonal bipyramid and support a concentration of *s* orbital participation in the equatorial bonds. Further, in considering the relative stability of isomers, accord is generally attained when ligand electronegativity variations are treated in the Y_nPX_{5-n} Series. Finally, the important area of isomeric rearrangement, which must be a major concern in the more elaborate computational schemes, reveals a notable degree of correspondence between the observed rate behavior and the energetics associated with assumed barrier states.

Excellent summaries of the general aspects of bonding are given in several articles *(1, 2, 3)*. We take up some of the more pertinent studies as they appeared chronologically, which somewhat parallels their order of mathematical complexity. However, we are not concerned with the mathematical elegence but rather the resulting bonding descriptions and their degree of applicability. Nonetheless, it is necessary to know the limits of the various methods, and an attempt is made to point these out.

0065–7719/80/0175–0001$20.00/1
© 1980 American Chemical Society

Table 1.1   Pentacovalent Bond Orbitals (4)

Structure	Symmetry	Species	σ Bonding	π Bonding
Trigonal bipyramid	D_{3h}	A_1'	s, d_{z^2}	
		A_2''	p_z	p_z
		E'	p_x, p_y $d_{xy}, d_{x^2-y^2}$	p_x, p_y $d_{xy}, d_{x^2-y^2}$
		E''		d_{xz}, d_{yz}
Square pyramid	C_{4v}	A_1	s	
			p_z d_{z^2}	p_z
		B_1	$d_{x^2-y^2}$	
		B_2		d_{xy}
		E	p_x, p_y $d_{xz}, d_{yz}{}^a$	p_x, p_y d_{xz}, d_{yz}

^{a} For an apical–basal angle of 90°, these d orbitals are pure π orbitals.

Models for Bonding in Pentacoordination

Directed Valence.   In an early look at bonding in a pentacoordinate molecule, Kimball (4), using group theoretical methods, classified likely hybrid orbital sets. The symmetries of the participating orbitals are listed in Table 1.1. However, most early calculations concerned with directed valence in pentacoordination assumed the $sp^3d_{z^2}$ hybrid set for σ bonds in trigonal bipyramids and the $sp^3d_{x^2-y^2}$ set for square pyramids. In an approximate treatment of PF_5 and PCl_5, using the former hybridization scheme, Cotton (5) found that the axial bonds are weaker than the equatorial bonds when the magnitude of the total overlap integrals is used as a criterion. Using a similar approach, Craig et al. (6) showed that the maximum in the radial wave function in an equatorial bond set of a trigonal bipyramid occurs at a smaller internuclear distance than that for an axial orbital set.

When several workers used Pauling's (7) criterion—i.e., that the bond strength is proportional to the projections of the angular parts of the atomic orbitals in the bond direction—they found it inadequate. Both Duffey (8) and Cotton (5), for example, have shown that its use leads to unrealistic predictions for the relative bond strengths of axial and equatorial linkages in PF_5 and PCl_5. Axial bonds became relatively weaker only as the limit of pure sp hybridization is approached for these bonds. Volkov et al. (9, 10) invoked this criterion in calculating the relative isomer stabilities for Cl_2PF_3. Their results suggested that the symmetric D_{3h} structure was the most stable of the three possible trigonal bipyramidal isomers; the C_s structure was next in stability, and the C_{2v} structure was least stable. At the time, the D_{3h} structure appeared reasonable, based on an electron-diffraction investigation (11). Later, however, the structure was shown to have C_{2v} symmetry (Volume I, p 112).

Simplified Molecular Orbital Approach.   Rundle (12–15) neglected *d* orbital participation in the σ framework in a first approximation for phosphorus(V) compounds and arrived at an orbital-deficient model of bonding. Three molecular orbitals are formed primarily from *p* orbitals in a description of axial bonding leading to three-center, four-electron bonding. One occupied molecular orbital is obtained involving all three centers, one occupied nonbonding orbital with most of the electron density on the two end atoms, and one empty antibonding orbital. Simply pictured, an axial bond order of

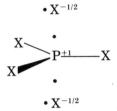

approximately one-half results, and there is a concentration of electron density at the end atoms. Thus, the model is of value in understanding not only the observance of longer axial bonds but the preference of electronegative groups for axial positions vs. equatorial sites.

Bartell and Hansen (*16*) extended this model by including contributions from the phosphorus 3*s* orbital to explain the observed increase in the ratio of the length of the axial P–F bonds to the length in the equatorial P–F bond(s) on going from PF_5 to CH_3PF_4 to $(CH_3)_2PF_3$ (Volume I, p 12). It is expected (*17, 18*) that because of the reduced electronegativity of the methyl ligands, *s* character will concentrate in the P–C bonds that are formed in equatorial sites. A corresponding reduction in *s* character of the P–F axial bonds produces a relative decrease in bond order. Although this simple molecular orbital approach is effective in accounting for the increase in the axial–equatorial bond length ratio with substitution, it implies a shortening of equatorial bonds, whereas, in fact, they are observed to increase in length. This, however, is not a valid shortcoming of an electron-deficient model, as we will show after looking at the valence shell electron-pair repulsion (VSEPR) model advanced so effectively by Gillespie (*19–24*).

Valence Shell Electron-Pair Repulsion Model. The VSEPR model deals principally with the stereochemistry of coordinated polyhedra. It may be regarded as a consequence of the Pauli exclusion principle, which states that there exists a mutual avoidance among electrons of the same spin. Since valence shells of atoms usually have equal numbers of electrons of opposite spin, which pair up upon formation of coordinated polyhedra, the theory becomes one of maximizing the average distance between localized electron pairs or minimizing the repulsive energy associated with their mutual avoidance. It is assumed that the repulsive interactions between the electron pairs in a valence shell may be represented by a potential function of the type used for repulsions among rare gas atoms. The latter involves a $1/r^n$ term, where *n* has a large value—e.g., 12, in the Lennard–Jones potential function. This leads to a short range interaction, which drops off rapidly as the distance between localized electron pairs increases.

Table 1.2.   Overlap Populations

Without 3d Orbitals

	ax P–F	eq P–F	ax P–Cl	eq P–Cl
PF_5	0.435	0.470	—	—
PF_4Cl	0.387	0.448	—	0.684
PF_3Cl_2	0.339	0.426	—	0.680
PF_2Cl_3	0.290	—	—	0.677
$PFCl_4$	0.265	—	0.432	0.660
PCl_5	—	—	0.416	0.647

For the problem at hand—that of five electron pairs around a central atom—the square pyramid and the trigonal bipyramid are equally probable when $n = \infty$ (hard sphere case). As the value of n is reduced, the trigonal bipyramid becomes more stable. For $n = 8$–12, the trigonal bipyramid is calculated to be about 8% more stable than the most stable square pyramid that has an axial–basal angle of about 100°. In an independent calculation, Zemann (25) and later Kepert (26), using a simple repulsion model with an apparent $1/r$ dependence, found a similar value for the apical–basal angle in the optimal square pyramid (104°), and Zemann (25) calculated an energy difference of 6.0 kcal/mol relative to the more stable trigonal bipyramid under an assumed M–X distance of 2.0Å. Kepert (26) also considered the effect of the variation of n in r^{-n} on the square pyramid geometry.

It is interesting to note that in the square pyramids found for $InCl_5^{2-}$ and $SbPh_5$ (Volume I, Table 2.1), the only two exceptions to the normally observed trigonal bipyramidal structure for nontransition elements with simple ligands, the axial–basal bond angle is close to predicted values—i.e., 103.9° (av) and 101.8° (av), respectively. We see, however, that VSEPR (19–24) cannot account for the appearance of the square pyramidal isomer for nontransition elements. Within a given structural type, the Gillespie model has achieved remarkable success in correlating distortions from idealized symmetries. In this sense, it is better than the simple molecular orbital arguments discussed above.

Although several corollaries may be formulated (19–21), we need only note that as electron pairs become more localized, brought about by the use of ligands of increasing electronegativity, repulsion with a neighboring orbital is reduced. If a molecule has several different types of ligands including nonbonding electron pairs, which would provide the largest most diffuse orbital, the geometry will be in accordance with a minimization of the total repulsion energy. Thus, a general ordering within a given valence shell regarding the decrease in repulsion effects is lone pair–lone pair > lone pair–bond pair > bond pair–bond pair. Because of the short range of the force

in P–X Bonds[a] (29)

With 3d Orbitals

ax P–F	eq P–F	ax P–Cl	eq P–Cl
0.486	0.527	—	—
0.380	0.431	—	1.082
0.269	0.331	—	1.028
0.158	—	—	0.985
0.071	—	0.700	0.920
—	—	0.658	0.855

law, repulsions in the idealized trigonal bipyramid at 90° will be far more important than those at 120°.

It is easy to understand then why the lone electron pair in the valence shell of $:SF_4$ occupies an equatorial site where it has more room or why, when five identical ligands are present, as in PF_5, the axial ligands with three nearest neighbors at 90° reach an equilibrium distance further from the phosphorus atom compared with the equatorial ligands that have two such neighbors. Moreover, since the flourine atom is the most electronegative, substituents will preferentially occupy equatorial sites with their larger electron pairs and will cause the more localized electron pairs to move away to minimize the overall repulsion. This line of reasoning (27) accounts for the increasing axial P–F distance and for the bend in the axial bond angle, away from the methyl substituents on going in the series from PF_5 to $(CH_3)_3PF_2$ (16). It also accounts for the fact that the equatorial ligands experience an increase in bond length with substitution, smaller than that observed in the axial bonds, an effect not readily interpreted by the simple molecular orbital approach (28).

Extended Hückel Molecular Orbital Calculations for the PCl_nF_{5-n} Series. That the latter effect can be accounted for in molecular orbital calculations, without getting too sophisticated, is apparent from an examination of overlap populations calculated by van der Voorn and Drago (29). They considered the bonding in the phosphorus chlorofluorides, PCl_nF_{5-n} using an extended Hückel approach, both with and without d orbitals in the basis set. These data are shown in Table 1.2.

Using the overlap population as an indication of relative bond strength, the P–F axial bonds are weaker than the P–F equatorial bonds in PF_5. Increasing chlorine atom substitution produces a general bond weakening in all the bonds but more so in axial than equatorial linkages. By analogy with the structural data in the $(CH_3)_nPF_{5-n}$ Series (Volume I, Table 2.1), better consistency is obtained when the calculation is done without 3d orbitals, in that the rate of decrease in implied bond strength is relatively smaller for the

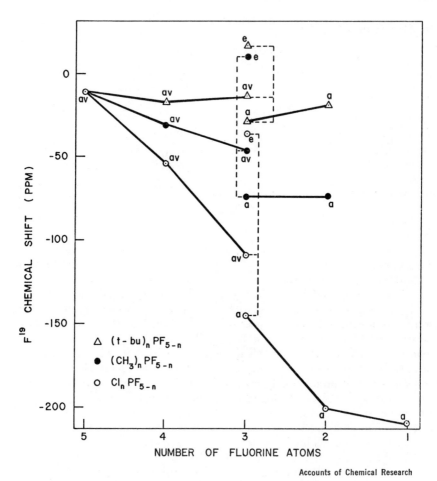

Accounts of Chemical Research

Figure 1.1. ^{19}F *chemical shifts relative to* CF_3CO_2H *(2)*

equatorial P–F bonds. Further, it was found in this calculation that phosphorus atom $3s$ character concentrates in the equatorial bonds, more so in the P–Cl bonds than in the P–F bonds, and hence concurs with Bartell's *(16, 28)* modification of Rundle's *(12–15)* orbital-deficient model.

Although there are a number of adjustable parameters in a Hückel type calculation, systematic calibration of the more important ones appears to have been achieved with use of the method by a variety of workers. Thus, one can have some confidence in the results, particularly if the method is applied to a series closely related in structure, as here. However, as a word of caution, a method not very different in rigor—i.e., a modified Wolfsberg–Helmholz approximation—gave *(30)* an energy difference of 66.3 kcal/mol, favoring the trigonal bipyramid relative to the square pyramid for PF_5. However, the focus of the calculation was not on attaining this quantity, so that parameterization may have followed a different course.

Although the Hückel calculations of van der Voorn and Drago (29) showed that the observed isomers in the chlorofluoride series agreed with their estimates of the most stable forms (for PF_4Cl their total energy estimates predicted the wrong isomer when no d orbitals were included but the correct one with d orbitals), this kind of treatment is not easily amenable to the calculation of angular distortions from idealized symmetry. On the other hand, the ease and reliability with which the VSEPR model (19–24) can predict geometrical distortions within a given framework remains its strongest feature in handling structural problems in nontransition element pentacoordination. (In an attempt to rationalize "why both the orbital model, without any explicit electron–electron interactions, and the Gillespie model, based on local repulsions, could both give such consistently useful predictions," Berry et. al. (30) commented that both do have a common basis, the Pauli exclusion principle. As a result, they state that "the short-range effective repulsive forces invoked in the Gillespie model do seem real and justified; they simply do

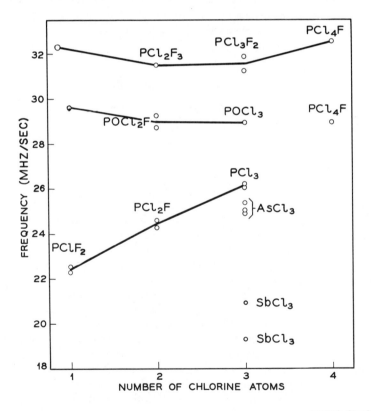

Inorganic Chemistry

Figure 1.2. ^{35}Cl *NQR frequencies (MHz) vs. number of chlorine atoms (31)*

not necessarily involve electrostatic repulsions between electrons." (This was embodied pretty well in the above approach taken in formulating (*19–21*) the VSEPR model.)

Correlation of Theory with Experimental Data in the PCl_nF_{5-n} Series. As chlorine substitution increases in the phosphorus chlorofluoride series, the implied lack of transmission of electron density, particularly to axial fluorine atoms (as noted by the pronounced deshielding of these atoms in the ^{19}F NMR (*2, 31, 32*) (Figure 1.1)), parallels closely the decrease in overlap population of the axial P–F bond (*29*). An almost linear relationship results between the ^{19}F chemical shifts and the overlap populations from Table 1.2, calculated without the use of d orbitals. It is also significant that the small variation in pure ^{35}Cl NQR frequencies (*31, 33*) for equatorial chlorine atoms (Figure 1.2) in this series is in essential agreement with the small changes present in equatorial P–Cl overlap populations (*29*). This appears to be corroborating evidence for the molecular orbital calculation (*29*) and the corresponding interpretation from the VSEPR model (*27*).

In terms of the orbital-deficient molecular orbital model (*12–15*), as previously argued (*2*), the resultant charges on the axial fluorine atoms in PF_5 decrease with substitution as the covalent contribution to the total bonding energy decreases in the series, leading to the structural transition from **A** to **B**.

A	**B**

Thus, the small gas-phase dipole moment observed for Cl_4PF (*34*) (0.21 D) is rationalized on this basis as well as on the ^{19}F chemical shift data.

The overlap populations (*29*), however, do not tell what the ionic terms contribute to bonding (*35*). This influence seems to make itself known when groups of greater electron-releasing ability than chlorine atoms are used. In the methyl and *tert*-butyl Series, R_nPF_{5-n}, transmission of the greater electron density supplied to the phosphorus atom should be facilitated and lead to more ionic fluorine atoms (*2*). In accord, the trend in ^{19}F shifts in Figure 1.1, for the Series $(tert\text{-}Bu)_nPF_{5-n}$, is more in line with group electronegativities.

Viewed in the context of the VSEPR model, replacement of ligands by those of reduced electronegativity enhances the electron repulsion between these less localized electron pairs and those in the P–F bonds, causing the latter to move further out. Both the center of the axial charge distribution and the fluorine atom center are involved in this movement. Thus, substitution in PF_5 may enhance the negative charge on the axial fluorines via the ionic term while the covalent contribution, which decreases as the P–F axial bond dissociates toward the atoms, tends to reduce the negative charge at fluorine.

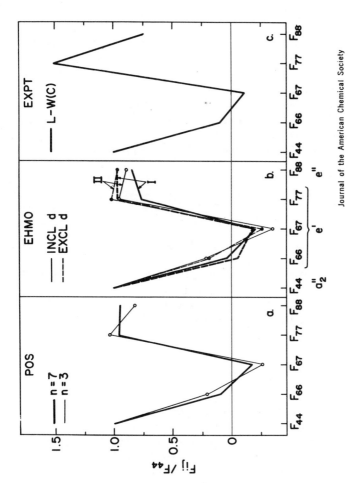

Journal of the American Chemical Society

Figure 1.3. Bending force constant profile for PF_5 (36). (a) Points-on-a sphere results for two values of the repulsion hardness parameter n. (b) Extended Hückel results for Set I (heavy lines) and Set II (light lines) parameters. (c) Levin-Wilt Field C. The Holmes Field B is similar to Levin-Wilt Field C.

Only with the stronger electropositive ligands does the ionic term manifest itself (2).

Force Field Comparison for PF_5 from the VSEPR and EHMO Treatments. A distinct advance in understanding the underlying significance of the VSEPR model (19–24) has been made by Bartell and Plato (36) in applying a points-on-a-sphere (POS) variant of this model, as well as an extended Hückel molecular orbital (EHMO) approach to the problem of molecular vibrations. It was shown that ratios of bending force constants, obtained from each of the treatments for PF_5, correlated well with the force field derived from an analysis of vibrational frequencies and electron-diffraction amplitude data (Volume I, Table 4.3, Wilt Field C; cf. Table 4.2, Set II). The degree of correspondence is apparent in Figure 1.3. In the EHMO calculations, Set I force constants correspond to the use of modified s and p orbital exponents of Clementi and Raimondi (37), and phosphorus $3d$ values from Bartell, Su, and Yow (38). Set II force constants are based on parameters reported by Hoffmann et al. (1).

The expressions for applying the POS approach are obtained easily from the assumed potential

$$V = K \sum_{k<l} \sum q_{kl}^{-n} \tag{1}$$

where q_{kl} is the distance between points k and l, which are constrained to move on the surface of a sphere of radius R; n is the repulsion coefficient, and K is a constant. The pentacoordinate atom is at the center of the sphere, and the five points represent the ligand atoms. The potential energy for small displacements in molecular vibrations under a quadratic approximation has been discussed for trigonal bipyramids of D_{3h} symmetry (Volume I, p 243). This potential

$$2V = \sum_{ij} F_{ij} S_i S_j \tag{2}$$

used in conjunction with the bending coordinates S_4, S_{6a}, S_{7a}, and S_{8a}, given in Volume I, p 242, and Equation 1 here, provide the necessary association between the POS model and deformation force constants. It has been shown (36) that force constant expressions for F_{ij} result by deforming the molecule along the symmetry coordinates S_i and S_j, expressing the resultant quantities q_{kl}^{-n} as a function of the angular displacements $\Delta\alpha$ and $\Delta\beta$, and relating these to S_i and S_j. These expressions, in units of $(2K/R^n)$, are

$$F_{44} = \left[3^{-n/2} \left(\frac{n}{4} \right) + 2^{-n/2} \left(\frac{n^2}{8} + \frac{n}{4} \right) \right] \tag{3}$$

$$F_{66} = \left[3^{-n/2} \left(\frac{n^2}{24} + \frac{n}{6} \right) \right] \tag{4}$$

$$F_{67} = -\left[2^{-n/2} \left(\frac{n}{\sqrt{24}} \right) \right] \tag{5}$$

$$F_{77} = \left[2^{-n/2} \left(\frac{n^2}{8} + \frac{n}{4} \right) + 2^{-n} \left(\frac{n}{6} \right) \right] \tag{6}$$

$$F_{88} = \left[2^{-n/2} \left(\frac{n^2}{8} + \frac{n}{4} \right) \right] \tag{7}$$

As seen in Figure 1.3 the POS results are not significantly altered for values of n of 3 or 7. In fact the results appear acceptable for $3 < n < 12$. However the POS interpretation suggests that the force law operating in the VSEPR model is substantially harder than coulombic. Coupled with the close agreement with the EHMO treatment (36) and experiment, the POS calculation confirms the Gillespie–Nyholm interpretation (19–24) that the mutual avoidance of occupied localized molecular orbitals is the proper basis for the VSEPR theory. A similar conclusion was noted by Berry et al. (30) (*see* p 7).

Polytopal Rearrangement

Exchange Barrier for PF_5 and PH_5. SQUARE PYRAMIDAL TRANSITION STATE. Until now we have discussed treatments concerned primarily with the elucidation of bonding in trigonal bipyramids. We now move into theoretical considerations of the square pyramid and the problem of small energy barriers that are responsible for facile intramolecular exchange. Associated with the latter behavior, the turnstile conformation will also be considered.

Table 1.3. **Structure and Activation Energy**[a] **of C_{4v} Pseudorotation Intermediate according to the Points-on-a-Sphere Model**[b] **(36)**

Hardness, n	Exact		Quadratic Approximation			
	θ_m	ΔE_B	F_{67}[c]	$\theta_m{}^{qa}$	$\Delta E_B{}^{qa}$	$\Delta E_B{}^{qa}/\Delta E_B$
2	103.6°	1.556	exact	104.0°	5.17	3.32
			zero	102.8°	14.79	9.50
4	102.4°	1.826	exact	102.2°	6.34	3.47
			zero	100.4°	15.24	8.35
7	100.8°	0.990	exact	99.7°	3.84	3.88
			zero	97.2°	8.40	8.48
10	99.5°	0.317	exact	97.6°	1.50	4.73
			zero	94.7°	3.30	10.43

[a] Energy in units of $(10^{-2}K/R^n)$.
[b] Data in the table are from Ref 36.
[c] Taken as its exact POS value or zero, retaining the exact values of F_{66} and F_{77}.

Journal of the American Chemical Society

After an appropriate coordinate transformation, the potential energy from Equation 1 for a square pyramidal structure, as shown by Bartell and Plato (36), in units of $(2K/R^n)$, is

$$V(C_{4v}) = 2[(2 \sin \theta/2)^{-n} + (2^{n/2} + 1)(2 \sin \theta)^{-n}] \qquad (8)$$

where θ is the apical–basal angle. Minimization of Equation 8 with respect to θ leads to the condition for the barrier structure in the Berry (39) process and the associated barrier energy, ΔE. These quantities are listed under "Exact" in Table 1.3 for different values of n.

To allow an estimation of a barrier energy from symmetry force constant values, normally available in a harmonic approximation, or from POS values, the E' quadratic force field was used (36).

$$2V = F_{66} S_{6a}{}^2 + F_{77} S_{7a}{}^2 + 2 F_{67} S_{6a} S_{7a} \qquad (9)$$

In reaching the square pyramid, the equatorial and axial bending coordinates are expressed as a function of the square pyramidal angle, θ

$$S_{6a} = \sqrt{6} \ [(2\pi/3) - \theta] \qquad (10)$$

$$S_{7a} = 2 \sqrt{3} \ [(\theta - \pi/2) + \arcsin(\cos^2 \theta)] \qquad (11)$$

The minimum energy condition for this approximation is found after substituting these expressions into Equation 9. The resulting terms using POS force constants are listed under "Quadratic Approximation" in Table 1.3. The values indicate the effect of neglecting anharmonicity to the extent that it is measured in the POS calculation and the effect of the neglect of the bend–bend interaction constant.

In practice, force constants derived from experimental data are used in the harmonic approximation to obtain θ_m and $\Delta E_B{}^{qa}$ values from Equation 9 and the minimization condition. Using the ratio in the last column, $\Delta E_B{}^{qa}/\Delta E_B$, the POS value, is obtained. Thus, using force constants from Volume I, Table 4.3, (Wilt Field C) in mdynes Å/rad^2 of 0.26, 0.31, and 4.17 for F_{66}, F_{67}, and F_{77}, respectively, Bartell and Plato (36) obtained $\theta_m = 97.0°$ and $\Delta E_B{}^{qa} = 22.1$ kcal/mol for PF$_5$. The POS values then result as $\theta_m = 98.1°$ (98.9°) and $\Delta E_B = 5.7$ kcal/mol (4.6 kcal/mol) for $n = 7$ (10). These values are close to that for PF$_5$ (3.8 kcal/mol) obtained by estimating anharmonicity relative to that observed (40) in VF$_5$ (see Table 1.4 for additional comparisons). The data in Table 1.3 should prove useful in bracketing intramolecular exchange barriers for other pentacoordinate molecules.

The POS estimation says that for a given n value, the bending force constants for different molecules should be in the ratio of K/R, and for values of n like 7 and 10, most of the dependence will be on the central atom–ligand bond radius. Relative to the value of 4.6 kcal/mol, determined by the POS model, the corresponding value for VF$_5$ is 1.8 kcal/mol, using a V–F bond distance of 1.71 Å. This is the same as the barrier value calculated from the

Table 1.4. Pseudorotational Barriers (kcal/mol)

Method	Ref	Orbital Type	PF_5	θ^a	PH_5	θ^a
Vibrational potential[b]	*40*		3.8		1.6	
Vibrational potential[c]	*41, 42*		2.8–3.3			
Hartee-Fock model	*45*	with *d*			3.9	103.57°[d]
Study[e]		no *d*			4.9	103.57°[d]
Extended Hückel	*1*	with *d*	1.4		2.1	
		no *d*	0.7	99.8°	2.3	99.8°
Ab initio	*3*	with *d*	4.8	101.34°	4.2	
		no *d*	8.5		4.8	
Ab initio	*44*				2.0	99.57°
CNDO	*46, 47*		3.5	105°		
Points-on-a-sphere	*36*		4.6	98.9°		
			5.7	98.1°		

[a] θ is the apical–basal angle in the square pyramidal barrier state. In the VSEPR model (*19–24*) it is about 100°; in a point charge (*25, 26*) calculation, $\theta = 104°$.

[b] *See* Volume I, p 267.

[c] Based on a two-dimensional anharmonic potential function (Volume I, p 273).

[d] An assumed value. No geometry search was made in this study.

[e] This label is used for ease of reference here. It is an ab initio study having a similar level of sophistication as that in Ref. *3*.

observed (*40*) anharmonic levels for VF_5 (Volume 1, p 269) using a one-dimensional treatment and is slightly higher than the barrier range of 1.2–1.5 kcal/mol obtained from a two-dimensional treatment (Volume I, p 273) (*41, 42*). Further, an ab initio calculation gave 1.9 kcal/mol for the exchange barrier for VF_5 (*43*). Thus a degree of confidence is provided in subsequent estimations for other members.

Additional calculations of the pseudorotational barrier are available for PF_5 and the hypothetical PH_5 molecule; some of them are based on more exact quantum mechanical calculations than we have described thus far. These values are listed in Table 1.4 (*1, 3, 36, 40–47*) and show good agreement with the values discussed.

In the ab initio study (*3*) of PF_5, geometry optimization yielded the following structures for the D_{3h} and C_{4v} representations.

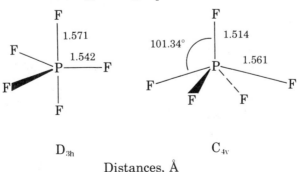

Distances, Å

The bond distances agree remarkably well with the known values of 1.534 Å (eq) and 1.577 Å (ax) for PF_5. The parameters for the C_{4v} barrier structure compare well with the established square pyramidal structures for $SbPh_5$ and $[InCl_5]^{2-}$ (Volume I, Table 2.1) in that all three exhibit shorter apical compared with basal bond distances and have apical–basal angles very near each other. It is felt that the θ value of 101.34° for the square pyramidal form of PF_5 represents a "best" value. This value is compared with others in Table 1.4.

In the more sophisticated and ab initio treatments (3, 44, 45), a lower barrier results when d orbitals are included in the basis set. Rauk, Allen, and Mislow (45) conclude that d functions serve only to remedy inadequacies in the basis set for PH_5. Strich and Veillard (3) show that their use implies effective participation in the bonding for the electronically more complex PF_5 molecule, giving a net transfer of 0.55 electron into the phosphorus d functions, with equal contributions from axial and equatorial fluorine atoms and π_{dp} bonding playing a significant role. Shown here are the results of population analysis from the two studies (3, 45), where d orbitals were included and one (in parentheses for PH_5) (44) using a 4-31 G basis set. Atomic charges and bond overlap populations are given for the D_{3h} structure and the assumed C_{4v} barrier state.

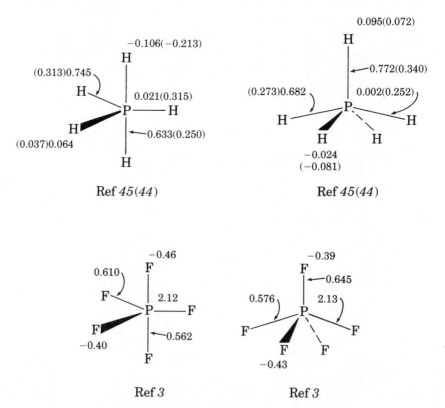

Ref 45(44)   Ref 45(44)

Ref 3   Ref 3

In agreement with conclusions from the other studies mentioned here (*1, 12–15, 29, 30, 44, 46, 47*) and some that are not covered in detail but should be consulted (*48–50*), the axial atoms of the trigonal bipyramid have greater negative charges and the axial bonds seem weaker compared to the corresponding quantities for equatorial bonds. Further, a correspondence between axial bonds (equatorial bonds) of the trigonal bipyramid and basal bonds (apical bonds) of the associated square pyramid is apparent. This conclusion was cited independently on the basis of electron-pair repulsion considerations (*2*) (Volume I, p 116).

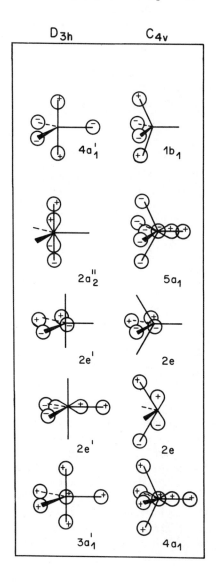

Journal of the
American Chemical Society

Figure 1.4. Symmetries of the higher occupied orbitals of PH_5 *(45)*

Journal of the
American Chemical Society

Figure 1.5. Correlation diagram for pseudorotation $(D_{3h} \rightarrow C_{4v})$ *of* PH_5 *(45)*

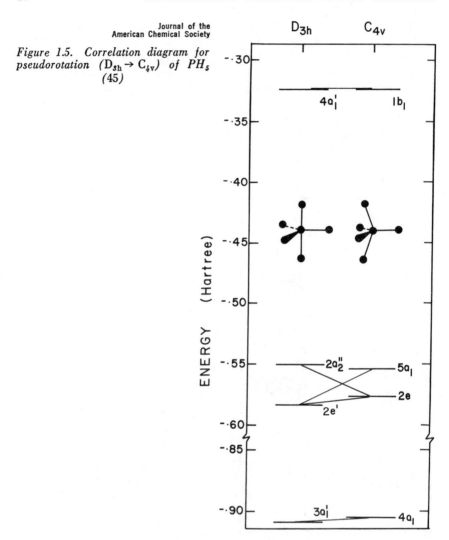

To aid in understanding the low barriers associated with pseudorotation for PF_5 and PH_5, the orbital description in Figure 1.4 and correlation diagram for the process $D_{3h} \rightarrow C_{4v}$ in Figure 1.5 are helpful. These are given from the study by Rauk et al. (45) for PH_5 but are similar for PF_5 with $2p\sigma$ atomic orbitals of the fluorine atoms substituted for the role of the hydrogen $1s$ orbitals (3). Only the highest occupied orbitals are shown for PH_5. The lower one shown (a_1') is *s*-like in character, the intermediate levels (e', a_2'') are *p*-like, and the upper level is *d*-like (a_1'). The latter is nonbonding in the absence of *d* orbitals, but it is bonding if *d* orbitals are included in the basis set. It is apparent from Figure 1.5 that little orbital change occurs during pseudorotation. Several groups have studied one (3) or more (1, 44, 46, 47, 48) steps along plausible sections of the potential energy surface connecting the two extreme structures. There seems to be general agreement that the energy of intermediate structures rises uniformly from the D_{3h} to the C_{4v} conformation with no indication of a saddle point in the potential (however, *see* Footnote c to Table 1.5). From these calculations, it becomes clear (48) that during the accompanying electronic redistribution, filled molecular orbitals mix with other filled ones, and empty ones mix with empty ones. The close similarity of structural and orbital makeup for the trigonal bipyramid and square pyramid, coupled with the existence of a continuum of structures between these two extremes situated along the Berry coordinate (Volume I, Figure 2.20), is strong evidence that the square pyramid is an adequate representation of the barrier in intramolecular exchange for PF_5.

TURNSTILE TRANSITION STATE. As a possible alternative, calculations involving the turnstile intermediate (Volume I, p 107) were done by several workers (1, 3, 44, 46, 47) using optimized geometries. Table 1.5 shows the barrier values by this route compared with the respective values these same workers obtained for the Berry process. The structures of the transition states are characterized by the angles α and β.

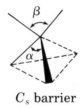

C_s barrier

The relatively large barriers associated with the turnstile mechanism represent a good argument against its implementation. In view of the closeness of orbitals between the D_{3h} and C_{4v} structures (Figure 1.5), it does not seem likely that activational entropy differences (*see* Footnote c, Table 1.5) could account for a kinetic stability to favor the turnstile path.

Isomer Stabilities in Derivative Phosphoranes. Several of the investigators (1, 3, 45) who have determined the barriers in Tables 1.4 and 1.5, have also calculated relative isomer energies for substituted phosphoranes. In general, there is agreement among the different workers as to the ordering of isomer stabilities within a given class. Inclusion of d orbitals tends to lower the barriers (exceptions are noted in Table 1.4) and to increase P–X bond orders (1). The ordering obtained by Rauk et al. (45) is shown as potential energy curves (Figures 1.6–1.9) and is compared with results of other studies in Table 1.6.

In the nonempirical model study of Rauk et al. (45) ligand electronegativity differences were introduced by increasing the nuclear charge of one or more of the protons of PH_5 to 1.1. This value corresponds to an electronegativity of about 2.9 on the Pauling scale, near that of the chlorine or nitrogen atoms. The absolute values in Table 1.6 are not expected to be

Table 1.5. Comparison of Barrier Energies for the Turnstile Mechanism and the Berry Exchange Mechanism, kcal/mol[a]

	Ref		Turnstile[b]			Berry
			ΔE	α	β	
Ab initio	3	PF_5	18.1	95°	85°	4.8
CNDO[c]	46, 47	PF_5	9.1	98°	85°	3.5
EHMO	1	PF_5	10.0	92.3°	85.6°	1.4
		PH_5	7.4	91.5°	87.2°	2.1
Ab initio	44	PH_5	10.1	92.6°	88.0°	2.0

[a] All calculations include d orbitals.
[b] Both the ab initio (3, 44) and CNDO (46, 47) calculations were based on a staggered transition state with respect to the orientation of the triad and pair. However in the EHMO (1) calculation only a 0.07 kcal/mol energy difference was found between the staggered and eclipsed conformations.
[c] The authors (46, 47) here argue that the exchange process may proceed by a turnstile mechanism that occupies a wide area on the potential energy surface (energy insensitivity to α and β within prescribed limits). The wide pass is entropy favored, whereas they regard the C_{4v} barrier at a saddle point on this potential surface. Other groups (1, 3) maintain that the high barriers rule out the turnstile process, and Hoffmann et al. (1) find a uniform energy change between the D_{3h} and either the C_{4v} (Berry) or C_s (turnstile) states, concluding that neither is an energy minimum.

especially significant with regard to a series containing chlorine or nitrogen atoms since, in these rather precise quantum mechanical calculations, the basis set was primarily proton-like. However, because of the sophistication of the ab initio approach, much significance is expected for the ordering of isomer stabilities. For the PH_2X_3 Series, this order is similar to that obtained by Strich (51) for PH_2F_3, based on an ab initio study, except for the last isomer entry—i.e., the trigonal bipyramid with both protons in axial positions. Here an unusually low value results.

If we invoke pseudorotational processes for intramolecular ligand exchange, the isomers identified by asterisks in Table 1.6 represent possible transition states (see Volume I, Figure 3.4 for pseudorotational processes for PY_2F_3 derivatives). The calculations show that where comparisons can be made, the relative order of barrier energies agrees with the general order of exchange rates observed for the different classes of phosphorus derivatives. The order of barrier energies from NMR data is $PF_5 < PYF_4 < PY_2F_3$ (p 41). Thus, it is pleasing that agreement between experiment and theory regarding important aspects of polytopal rearrangements is achieved (cf. Table III of Ref 70).

Further understanding of detailed structures of proposed exchange intermediates will most profitably result from experiments that are designed to detect and to isolate them—always an elusive undertaking—and from additional structural characterization of stable pentacoordinated species. The difficulty in establishing a mode, much less a mechanism of positional exchange has received emphasis by Musher (52). Dalton (53) has examined the possi-

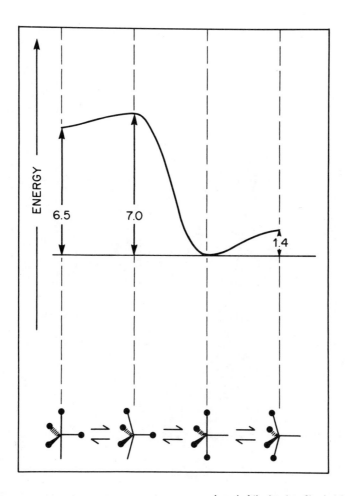

Figure 1.6. Potential energy curve connecting 'all trigonal bipyramidal and square pyramidal structures of PHX$_4$ (45). The more electronegative X is represented by a black circle. The numbers are in kcal/mol.

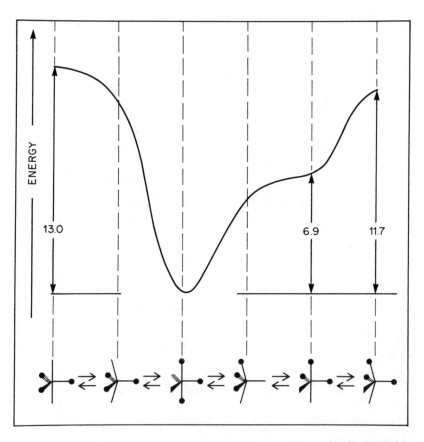

Figure 1.7. Potential energy curve connecting all trigonal bipyramidal and square pyramidal structures of PH_2X_3 *(45). The more electronegative* X *is represented by a black circle. The numbers are in kcal/mol.*

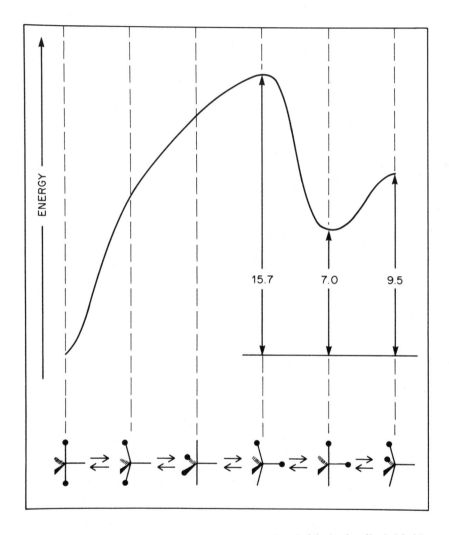

Figure 1.8. *Potential energy curve connecting all trigonal bipyramidal and square pyramidal structures of* PH_3X_2 *(45). The more electronegative X is represented by a black circle. The values are in kcal/mol.*

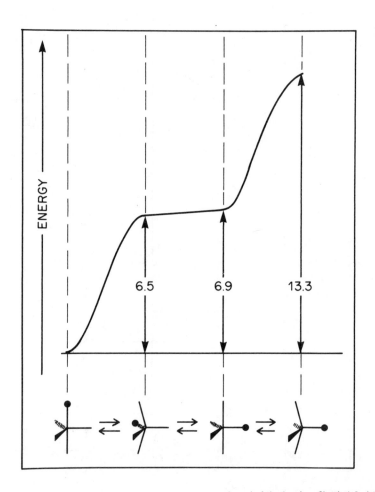

Figure 1.9. Potential energy curve connecting all trigonal bipyramidal and square pyramidal structures of PH_4X (45). The more electronegative X is represented by a black circle. The numbers are in kcal/mol.

Table 1.6. Isomer Energies Relative to Ground-State Trigonal Bipyramids of PH_nX_{5-n} (kcal/mol)[a]

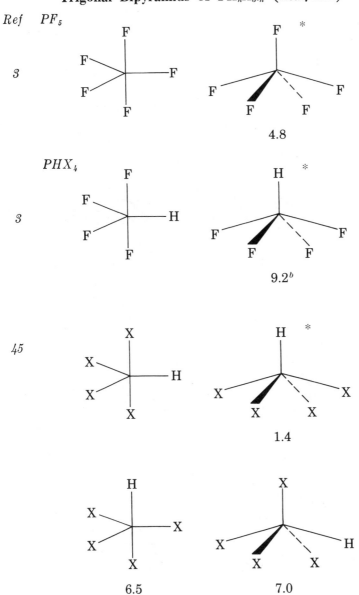

Ref PF_5

3

4.8

PHX_4

3

9.2^b

45

1.4

6.5   7.0

Table 1.6.   Continued

Ref PH_2X_3

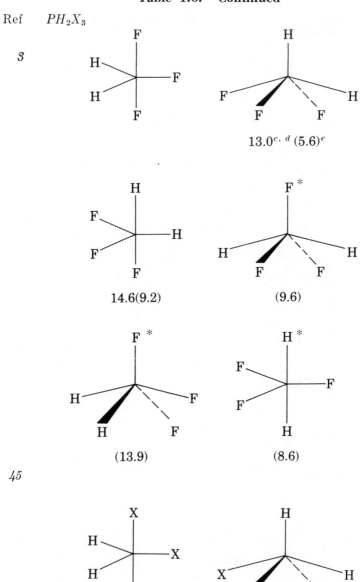

3

13.0[c, d] (5.6)[e]

14.6(9.2)   (9.6)

(13.9)   (8.6)

45

5.5

Table 1.6.   Continued

Ref PH_2X_3

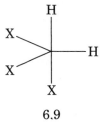

6.9

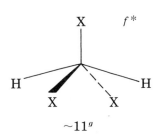

f^*

$\sim 11^g$

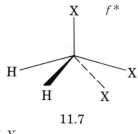

f^*

11.7

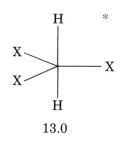

$*$

13.0

PH_3X_2

45

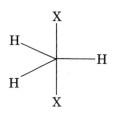

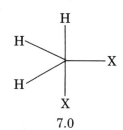

7.0

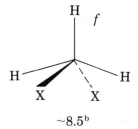

f

$\sim 8.5^b$

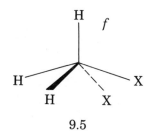

f

9.5

Table 1.6.   Continued

Ref   *PH$_3$X$_2$*

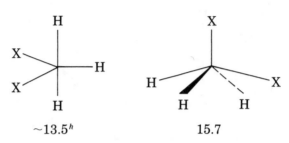

~13.5^h 15.7

3   *PH$_4$X*

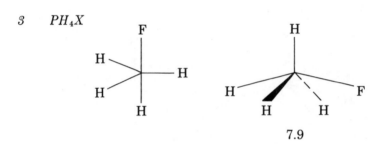

7.9

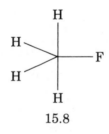

15.8

45

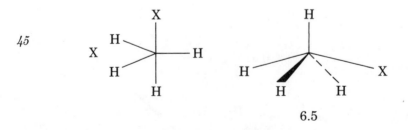

6.5

Table 1.6.   Continued

Ref   *PH₄X*

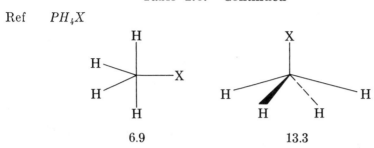

6.9   13.3

[a] Values from Refs *1* and *3* are based on calculations including *d* orbitals. For the substituted phosphoranes, Ref *45* does not indicate whether *d* orbitals were included or not. Asterisks identify isomers that may represent ligand exchange barriers in pseudorotational processes.

[b] The value reported in Ref *3* without the use of *d* orbitals was 7.5 kcal/mol in contrast to the usual increase observed.

[c] As in Footnote *b*, the value here was lowered to 10.2 kcal/mol when *d* orbitals were not used (*3*).

[d] In the CNDO calculation (Refs *46, 47*) the value was 6.6 kcal/mol. This compares with the lower value of 3.5 kcal/mol that these authors reported for the square pyramid associated with PF₅.

[e] Values in parentheses are for PH_2F_3 isomers obtained by Strich (*51*) from an ab initio treatment.

[f] The order of isomer stabilities predicted from the EHMO method of Hoffmann et al. (*1*) agrees with the ordering in this table except for the interchange of these square pyramidal pairs. However, trigonal bipyramid and square pyramid orders were compared as independent sets, and no values were given.

[g] Approximate value estimated from Figure 1.7.

[h] Approximate value estimated from Figure 1.8.

ble effects of quantum mechanical tunneling on rovibronic levels for a non-riged molecule, such as PF₅, and calculated characteristic patterns that could be searched for in the microwave region. Because of the greater nonrigidity associated with VF₅, it would make a better candidate for this type of experiment (Volume 1, p 275).

Inhibition of Exchange by π Systems.   A further aspect of substituent behavior has been considered carefully by Hoffmann et al. (*1*) and by Strich and Veillard (*3*) regarding inhibition of intramolecular exchange in amino- and thiofluorophosphorane derivatives caused by π bonding effects. Relative to the energy difference of 9.2 kcal/mol between the ground-state trigonal bipyramid and lowest energy square pyramidal isomer for HPF₄ (Table 1.6), this energy difference increases substantially when the substituent bears a single π system. In NH₂PF₄, for example, with *d* functions on phosphorus included but none on nitrogen and assuming a planar nitrogen atom with the amino group oriented in an axial plane of the trigonal bipyramid, the difference is 18.0 kcal/mol (*3*). Inclusion of *d* functions on nitrogen has a stabilizing effect. Thus, the energy of the trigonal pyramid, referred to above, is lowered by 6.4 kcal/mol with their use (*3*).

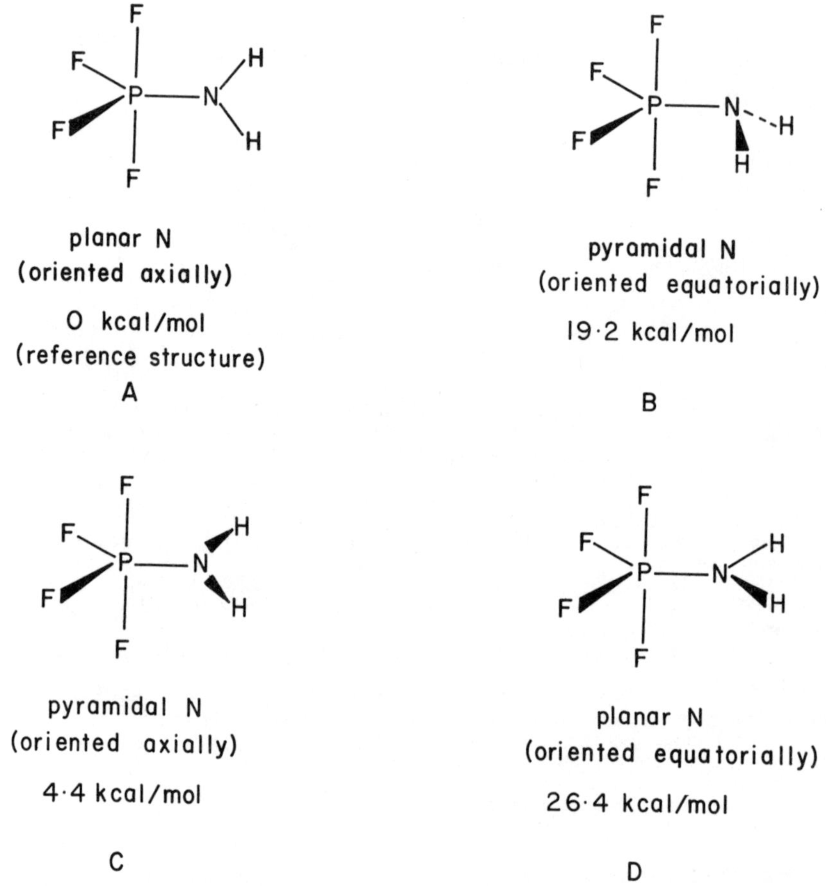

planar N
(oriented axially)

O kcal/mol
(reference structure)

A

pyramidal N
(oriented equatorially)

19·2 kcal/mol

B

pyramidal N
(oriented axially)

4·4 kcal/mol

C

planar N
(oriented equatorially)

26·4 kcal/mol

D

Figure 1.10.   Conformations for NH_2PF_4. Energy values are those calculated by Strich and Veillard (3), which included d functions on both the
nitrogen and phosphorus atoms.

In the process of rotation about the P–N bond, it is calculated (3) that
the conformation with the amino group oriented at 90° to that above (but
with pyramidal nitrogen) is 19.2 kcal/mol less stable (Figure 1.10). Relative
energies of other conformations are shown in Figure 1.10. Consideration of
rotation involving a π system in an apical position of the square pyramid
reveals an insignificant energy barrier. For PH_4SH, a 0.2 kcal/mol difference
is found for staggered and eclipsed structures (3). Thus, the slow exchange
limit seen experimentally via low-temperature NMR measurements for amino-
and alkylthiotetrafluorophosphoranes (Volume I, pp 103 and 145), in contrast to rapid exchange deduced for other members of the XPF₄ Series (Volume I, p 102), appears validated by the high barriers to pseudorotation ob-

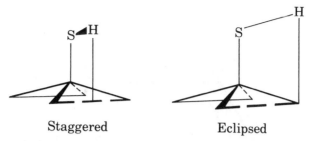

Staggered   Eclipsed

tained from calculations of the type above. The lowest energy exchange path, which couples a Berry process with bond rotation, suggests that the amino group in H_2NPF_4 is planar in the more stable axial orientation of the trigonal bipyramid but pyramidal as rotation continues (3). Inclusion of d functions at nitrogen stabilizes the pyramidal structure much more than the planar arrangement (3).

Compared with experimental estimates on related molecules, the computed value of 18 kcal/mol for the exchange barrier in H_2NPF_4 seems high. NMR data on positional exchange in $(CH_3)_2NPF_4$ (54) gave a barrier value of 9.5 kcal/mol (55). An estimated P–N rotational barrier of 11.1 kcal/mol in $(NH_2)_2PF_3$ was calculated (55) from ^{19}F line shape analysis, which supported uncorrelated rotation of the amino groups. The use of a limited basis set and lack of geometry optimization in the calculations of Strich and Veillard (3) may account for part of the differences. Nevertheless, the data verify the fact that P–N bond rotation makes a major contribution to the Berry process in aminofluorophosphoranes (R_2NPF_4), as pointed out by Muetterties et al. (56).

Perhaps exploration of additional points on the potential energy surface will provide better agreement. Since hindered rotation limits the exchange process, a more realistic calculation might be provided by starting at a point somewhat removed from the idealized trigonal bipyramid, A (Figure 1.10), where the NH_2 group that contains planar nitrogen is canted relative to the axial direction. With some degree of bond bending along the Berry coordinate also present, this ground-state representation would require less energy to rearrange to a square pyramidal barrier state. The latter is in agreement with reported x-ray structures that contain amino substituents (Volume I, Entries **5** and **14**, Table 2.3).

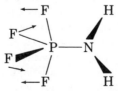

In their treatment of polytopal rearrangements, Hoffmann et al. (1) have generalized conformational preferences for π bonding substituents. It is argued that π donors will favor equatorial positions of a trigonal bipyramid, π acceptors, axial positions. If substituents that contain a single π system, similar to the examples cited above, are located in equatorial positions, they

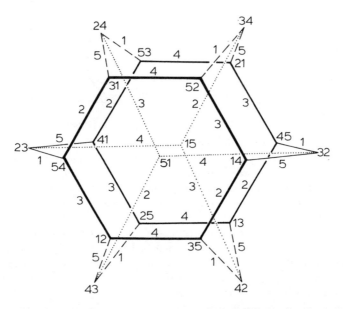

Journal of the American Chemical Society

*Figure 1.11. Topological diagram (57) for pseudorotations summarizing isomerization (heavy solid lines) and epimerization processes (light solid lines) for compound **I** (see Figures 1.14 and 1.15). Higher energy epimerizations are dashed lines, and forbidden isomerizations for **I** are dotted lines. The numbers refer to the axial positions in the TP or apical position in the SP as shown in Figures 1.14 and 1.15. The placement of basal ligands in the SP is established by the pairs of numbers identifying the two connecting TP's. These will be trans pairs in the SP (70).*

will prefer to have their acceptor orbitals perpendicular to the equatorial plane or their donor orbitals in that plane. For square pyramids, π donors will favor apical positions, and π acceptors will favor basal sites.

Topological Representations. In discussing complex polytopal rearrangements that may involve more than one isomeric conversion, a formal process is necessary to avoid ambiguity in describing mechanistic pathways. These interconversions have been summarized most conveniently in terms of topological representations (57) such as the one shown in Figure 1.11. The edges represent square pyramidal states under the Berry hypothesis. Topological maps represent the turnstile process (Volume I, p 107) equally well since this process is permutationally indistinguishable from the Berry process. If turnstile rotation is invoked, the edges represent the transition state peculiar to that mechanism. However, since most of the literature of interest here has been interpreted in terms of the Berry pseudorotational process, and both theoretical estimates (p 18) and solid-state structural distortion (Volume I, pp 34 and 42) favor the Berry process, we do not consider the turnstile process further (Volume I, p 110).

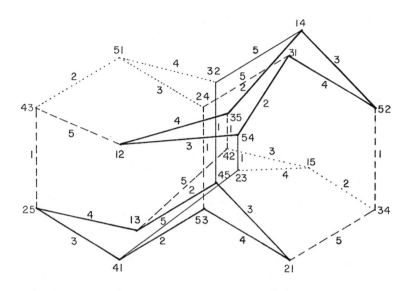

*Figure 1.12. Topological diagram (59, 60) comparable with Figure 1.11. With reference to the pseudorotational processes for **I** (see Figures 1.14 and 1.15), the different types of lines in the present figure have the same meaning as described in the caption for Figure 1.11.*

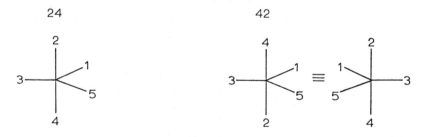

Figure 1.13. Illustration of the convention for numbering ligands in a topological representation

The topological representation shown in Figure 1.11 has been used widely (57, 58) in relating isomer configurations that result from pseudorotational processes. An equivalent representation (59, 60) but less symmetrical in appearance is shown in Figure 1.12. In using these representations we number the axial positions by starting with the upper one first. The equatorial substituents acquire numbers so that they increase counterclockwise (61). For example, Epimers 24 and 42 have the ligand numbering illustrated in Figure 1.13. Other topological representations (46, 47, 62–69) have proved useful as an aid in determining the least number of isomerizations encountered in a minimum energy pathway. The literature should be consulted for further appreciation of this subject. For our purposes, we refer to the graph in Figure 1.11.

The relationship between a postulated square pyramidal (SP) and trigonal bipyramidal (TP) transition state can be seen more clearly by examining the complete interconversions accompanying the isomerization (Figure 1.14) and epimerization (Figure 1.15) of the addition product of 1 (59). With reference to the topological diagram in Figure 1.11, the highest energy TP conformations that must be traversed for the isomerization process are Isomers 35 and 52, while the highest energy SPs are 2 and 3. For the epimerization of 31, a TP of higher energy containing a diequatorial ring must be reached. This is Isomer 23, connecting the isomerization routes. The latter are shown by heavy solid lines in Figure 1.11. The high energy SP isomer for this process is 1. If effective comparisons are to be made, it becomes necessary to devise complementary scales that will establish an order of isomer energies involving both trigonal bipyramidal and square pyramidal conformations.

Isomer Energies Based on a Model Approximation. In Chapter 2, we discuss mechanisms for important classes of phosphorus reactions that seem to involve one or more five-coordinated intermediates. Based on the accumulated data presented here and in the preceding volume, an evaluation of the structural makeup and relative energy of the reaction intermediates is feasible. In a quasi-quantitative approach (70), the author has developed a model, a first approximation, to accomplish the foregoing.

The model allows the estimation of the relative stability of all TP and SP isomers of a given pentacoordinate phosphorus compound. The Berry mechanism (Volume I, p 106) is adopted as the mechanism that controls postulated pseudorotations. This provides the connection between intervening SP isomers formed on the way to an interconverted TP (cf. Figure 1.11).

A series of acyclic phosphoranes free of ring strain and steric factors was chosen to calibrate the model internally. The development leads to apicophilicity scales (Table 1.7), which are constructed for the TP and SP with the additional constraints that they have the same range and correlate directly with each other as a function of ligand electronegativity.

A major achievement (70) of the use of this scale (Table 1.7) is the close correspondence of relative isomer energies for the Series PH_nF_{5-n} with those reported by Rauk et al. (45), based on nonempirical molecular orbital methods. Not only is the order of isomer energies the same with one excep-

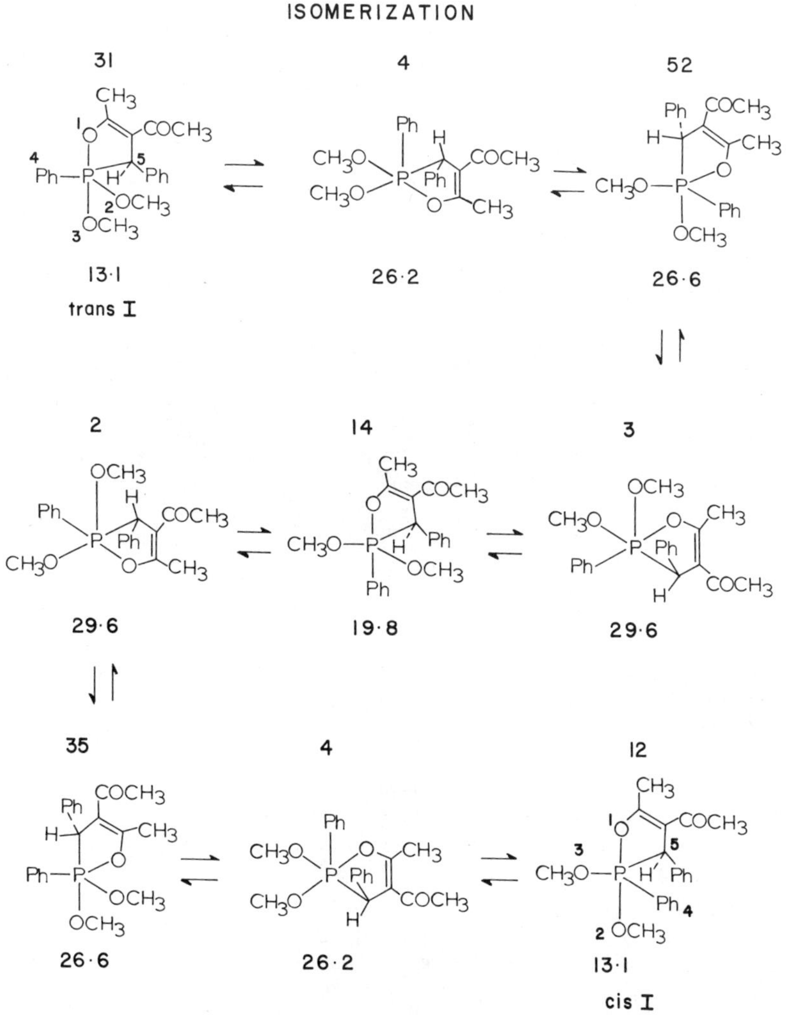

ISOMERIZATION

*Figure 1.14 Intramolecular isomerization of the adduct of dimethylphenylphosphonite and benzylideneacetylacetone **I** via the low energy pseudorotational pathway (59). The numbers above each structure identify the isomer on the topological diagram shown in Figure 1.11. The numbers below each structure give the relative isomer energy in kcal/mol via the terms in Tables 1.5 and 1.7–1.9 (70).*

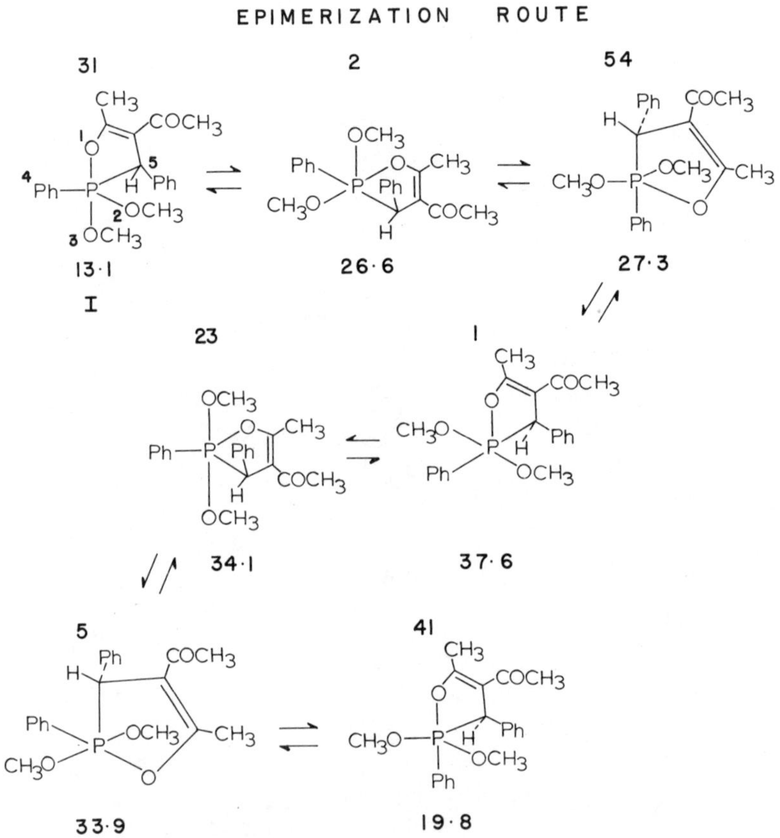

EPIMERIZATION ROUTE

Figure 1.15. Part of the intramolecular epimerization route of the adduct of dimethyl phenylphosphonite and benzylideneacetylacetone I via the low-energy pseudorotational pathway (59). The numbers above each structure identify the isomer on the topological diagram shown in Figure 1.11. The numbers below each structure give the relative isomer energy in kcal/mol via the entries in Tables 1.5 and 1.7–1.9 (70).

Table 1.7.   Relative Values of the Element Effect for Use in Estimating Isomer Energies (kcal/mol)[a] (*70*)

| X^c | TP | | A^b | | SP | | $-A^b$ | |
	eq	ax	ax-eq	$-A^d$	ap	bas	bas-ap	A^d
C	2	0.0	7.0	7.0	0.0	0.0	4.0	4.0	0.0
H	2.5	1.3	4.7	3.4	3.6	2.3	2.2	−0.1	4.1
Ph	2.7	1.7	3.9	2.2	4.8	3.0	1.7	−1.3	5.3
N, Cl	3.0	2.2	2.8	0.8	6.2	3.8	1.2	−2.6	6.6
O	3.5	2.8	1.3	−1.3	8.3	5.1	0.4	−4.7	8.7
F	4.0	3.0	0.0	−3.0	10.0	6.0	0.0	−6.0	10.0

[a] To the sum of these element values for a SP, 7.0 kcal/mol must be added to obtain the energy of an acyclic SP relative to a TP isomer. If steric and/or ring strain effects are present, these terms must be included. Ring strain values are in Table 1.8, and steric effects are in Tables 1.9 and 1.10.
[b] Relative values for apicophilicity (*A*) in a TP (ax–eq) or SP (ap–bas). The smaller or more negative values of *A* indicate a greater stabilization for a particular element in an apical position of either a TP or SP.
[c] Electronegativity scale.
[d] Apicophilicity scale relative to C equal to 0.

Journal of the American Chemical Society

tion, the second and third isomer for PH_4X, but the values averaged to within ±0.5 kcal/mol of each other over the entire series listed in Table 1.6 for Ref *45* (p 23).

For cyclic and sterically hindered phosphoranes, it is possible to arrive at reasonable ring and steric strain terms (Tables 1.8–1.10) by examining $\Delta G^{\neq}$ values for intramolecular ligand exchange for a series of related derivatives in which these effects seem to vary systematically. Application of the resulting model (*70*) reproduces measured $\Delta G^{\neq}$ for ligand rearrangements on cyclic and acyclic derivatives to within ±1.5 kcal/mol. (Table 1.11). It is hoped that others will extend the treatment as new data become available and thus improve its usefulness.

The following corrections to the article by Holmes (*70*) should be noted. Table IV: Entry 9, TP C 16.4, SP B 16.0, SP F 17.0; Entry 11, SP 9.1; Entry 15, TP 14.5, SP 17.5; Entry 16, TP 13.5; Entry 18, SP 22.1; Entry 25, SP 13.0; Entry 26, SP 18.1; Entry 27, SP 17.3; Entries 34 and 35, SP 14.7; Entry 41, TP 10.0; Entry 43, SP 17.9. Table V, delete Entry $R_2NC(CH_3)_2$. Table IX, successive entries under the column labeled XI should read 44.8, 47.8, 50.8, and 56.8, and Footnote *c* deleted. The first sentence on p 445 should read, "The SP XI is traversed on the way to the transition state X." Below Isomers 52 and 35 of Figure 2, the number should be 26.6 and Formula f on p 442 should have an R appended to the ring substituent. These corrections have been incorporated in Table 1.11 presented here.

Examination of the utility of the numerical scales (Tables 1.7–1.10) in predicting TP–SP isomer stability shows an excellent correlation with ground-state geometries obtained from x-ray analysis and NMR measurements. The

Table 1.8.   Relative Values of Ring Strain for Use in Estimating Isomer Energies (kcal/mol) (70)

Ring Structure	TP		SP	
	ax–eq	eq–eq[a]	bas–bas	ap–bas
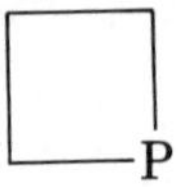	8	24	2	16
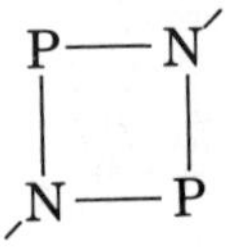	2		1	
	2	15	1	9
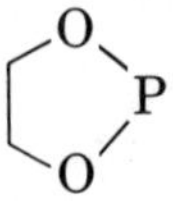	2	18	1	14
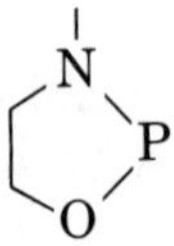	2	21[c]	1	14
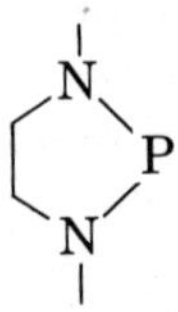	2	24[c]	1	14
	2	15	1	9
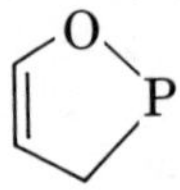	2	21	1	13
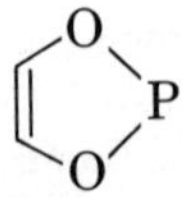	2	23	1	15

[a] Values for rings spanning apical–basal sites of a SP are approxiately 0.6 of the values for strain estimates for rings positioned in diequatorial sites of a TP.
[b] Present in the bisbiphenylylene derivatives (**43–52**) of Table 1.11.
[c] Variation in the strain energy for the oxaza rings in the TP reflects a 3 kcal/mol loss in π bonding for this orientation of a nitrogen atom.

Journal of the American Chemical Society

trend in relative ground-state isomer stabilities for cyclic phosphoranes obtained from the model, $\Delta(SP-TP)$, and that calculated from x-ray data—i.e., percent displacement along the TP–SP (RP) coordinate (Volume I, p 45) are in general directly comparable *(70)*. The appearance of the two isomers, **E** and **F** (Volume I, p 221), in the ratio of 2.3 to 1, respectively, at $-100°C$, but undergoing rapid interconversion at 25°C based on ^1H NMR, is explicable in terms of a substantial barrier between them. The relative energies (kcal/mol) for the isomeric sequence in Volume I, Figure 3.18, **31 ⇄ 2 ⇄ 54 ⇄ 1 ⇄ 23** are, respectively, 27.7, 34.7, 35.9, 38.7, and 29.7 giving an average barrier $\Delta G^{\ddagger}$ of 10 kcal/mol between the ground states **E** and **F** that are separated by about 2 kcal/mol *(70)*.

To facilitate the accessibility of the approach, a simple FORTRAN program (ISOMER 2) is appended, which incorporates all the element effects, ring strain values, and steric factors found in Table 1.7–1.10 (summarized from Ref *70*) as well as additional entries based on subsequent data. The latter entries along with their associated values in kcal/mol (in parentheses) are the steric terms, Ph–P–CH$_2$Ph (3), R$_2$N–P–OR (2), (CH$_3$)$_2$Ċ–P–CH$_2$Ph (3); a π effect for sulfur, SH or SR, (2); the ring strain terms

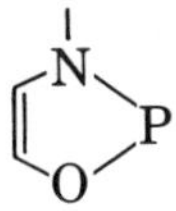

$\overset{|}{N}$ ring P equated to $\overset{|}{N}$ ring P,

$\begin{matrix} C-N' \\ | \quad | \\ N-P \end{matrix}$ equated to $\begin{matrix} P-N' \\ | \quad | \\ N-P \end{matrix}$,

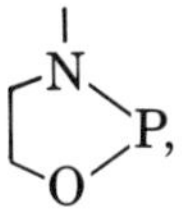

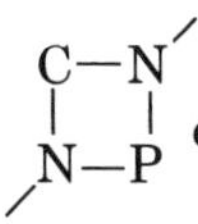

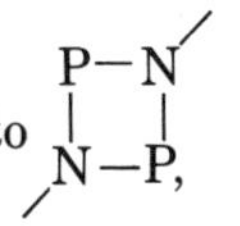

and

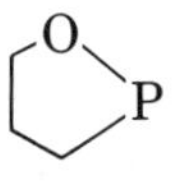

P with values interpolated between P and P.

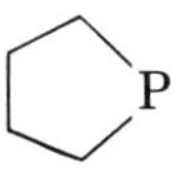

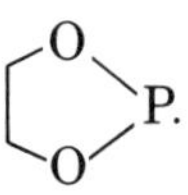

The π effect for a cyclic nitrogen atom, R–Ṅ–P, was decreased to 2 kcal/mol from the 4 kcal/mol initially assigned (cf. p 442 of Ref *70*). The new substituents, CF$_3$ and SR (SH), are assigned electronegativities of 4.0 and 2.5, respectively. Less calibrated steric terms for *tert*-butyl groups are included: (CH$_3$)$_3$C–P–C(CH$_3$)$_3$ (8), (CH$_3$)$_3$C–P–Ph (7), (CH$_3$)$_3$C–P–CH$_2$Ph (6), and (CH$_3$)$_3$C–P–CH$_3$ (5). In addition, certain substituent effects found particularly applicable to pentacoordinated intermediates postulated in reaction mechanisms that are discussed in Chapter 2 are defined. A complete list of all substituent effects is given on p 89.

ISOMER 2 calculates isomer energies for all possible trigonal bipyramidal and square pyramidal isomers for any five given substituents on phosphorus.

Table 1.9. Relative Values of Steric Factors for Use in Estimating Isomer Energies[a] (70)

Steric Interaction[b] (R = CH$_3$, C$_2$H$_5$)	Steric Factor, kcal/mol	Steric Interaction[b]	Steric Factor, kcal/mol
R$_2$N–P–NR$_2$	6.5	CH$_3$$\overset{\vert}{N}$–P–CH$_3$	2.0
		PhO–P–$\overset{\vert}{C}$HPh	2.0
Ph–P–Ph	6.0	RO–P–$\overset{\vert}{C}$HPh	2.0
Ph–P–$\overset{\vert}{C}$(CH$_3$)$_2$	6.0	CH$_3$$\overset{\vert}{N}$–P–Cl	2.0
		F–P–Ph	2.0
Ph–P–$\overset{\vert}{C}$HPh	5.0		
Ph–N–P–C$_2$H$_5$	5.0	RO–P–Ph	1.0
(CH$_3$)$_2$N–P–$\overset{\vert}{C}$(CH$_3$)$_2$	4.0	RO–P–OR	1.0

[a] For application to axial–equatorial orientations in the TP and both apical–basal and basal–basal cis ligand orientations in the SP. All interactions for a particular isomer should be summed to obtain the steric destabilization. The only exception that arises is when two identical steric interactions are present that involve a single apical substituent and this substituent (not a member of a ring) has no other steric interactions. In this case, a value is taken for these two interactions that equals 1.5 times that for a single interaction. Entry 32 of Table 1.11 represents this type of exception.

[b] The P–C– and P–N– bonds are part of ring systems.

[c] These represent steric terms between substituents of the ring that are applied only when these rings are oriented in axial–equatorial positions of a TP.

[d] This is not a steric term but represents a loss in π bonding and is applied in an additive sense when these groups are in locations other than the equatorial site of a TP.

All contributions to the total energy of each isomer are tabulated (i.e., contributions from ring strain, substituent effect, steric interaction, etc.) and printed. A table of isomers in order of decreasing energy is also printed. Sample calculations are given later in this chapter for entries 8 and 17 of Table 1.11. Sample calculations are explicitly illustrated in Ref 70 for Entries 8 and 16 of Table 1.11.

Additional studies appearing since the publication of the model approach (70), which provide activation energies governing fluxional behavior, concern two principal types of phosphoranes: oxazaspirocyclics (Volume I, p

Table 1.9. Continued

Steric Interaction[b] *(R = CH_3, C_2H_5)*	*Steric Factor, kcal/mol*	*Steric Interaction*[b]	*Steric Factor, kcal/mol*
R–P–Ph	4.0	RO–P–OPh	1.0
OR–P–$\overset{\textstyle\vert}{C}(CH_3)_2$	4.0	$CH_3\overset{\textstyle\vert}{N}$–P–OCH$(CF_3)_2$	1.0
$CH_3\overset{\textstyle\vert}{N}$–P–Ph	4.0	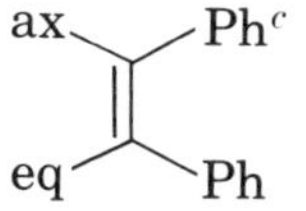	3.0
PhO–P–$\overset{\textstyle\vert}{C}(CH_3)_2$	3.0	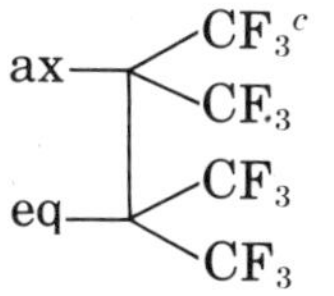	
$CH_3\overset{\textstyle\vert}{N}$–P–$C_2H_5$	3.0		2.0
$CH_3\overset{\textstyle\vert}{N}$–P–OPh	3.0		
		NR$_2$, NH$_2$[d]	6.0
		R$\overset{\textstyle\vert}{N}$–P[e]	4.0

(structures shown in right column: ax— Ph[c] / eq— Ph at 3.0; ax—CF$_3$[c], —CF$_3$, eq—CF$_3$, —CF$_3$ at 2.0)

[e] This is the same type of term as in Footnote *d* but refers to an NR group as part of a four- or five-membered ring. It is applied in the same manner as the term in Footnote *d*. If the ring is oriented diequatorially, it might be expected that this term should also be applied since the equatorial nitrogen atom is not properly oriented for effective π bonding, at least according to extended Hückel calculations (*1*). However, the resultant loss in π bonding must not be as great as that encountered when a nitrogen atom is located in other than an equatorial site of a TP (*see* Footnote *c* to Table 1.8). X-ray structures of phosphoranes containing Me₂N groups in equatorial sites show these groups in planes rotated 20°–30° from the axial direction (Entries **5** and **14**, Volume I, Table 2.3 and (NMe₂)₃PF₂, Table 2.1). Presumably, changes in π bonding effects for oxygen as a function of TP or SP site occupancy are even less important than those for nitrogen.

Journal of the American Chemical Society

168) and acyclic derivatives that contain trifluoromethyl groups (Volume I, p 120). Application of the model (*70*) leading to calculated free energies of activation for epimerization of the oxaza compounds is straightforward. Use of the FORTRAN program results in the comparison of calculated and experimentally determined $\Delta G^{\ddagger}$ values given in Table 1.12. The agreement between the two sets of $\Delta G^{\ddagger}$ values is ± 2.6 kcal/mol. In arriving at these values, the substituent effect for the O–C (ring) rather than that for the oxygen atom was applied (*see* Table 2.1) as well as the revised π effect for a ring-containing nitrogen atom (*see* p 37). If the values of these effects as applied in the original

Table 1.10.   Relative Values of Steric Effects for Substituents in Bisbiphenylylene Derivatives [a, b, c] (70)

R	Steric Factor, kcal/mol	R	Steric Factor, kcal/mol
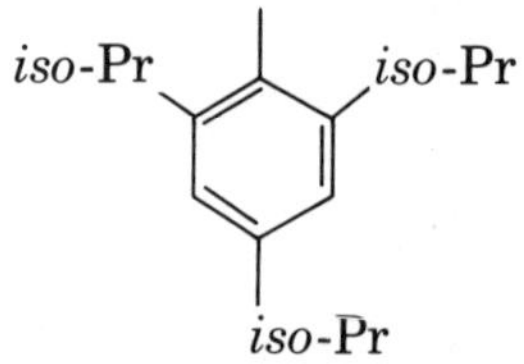 iso-Pr, iso-Pr, iso-Pr	10	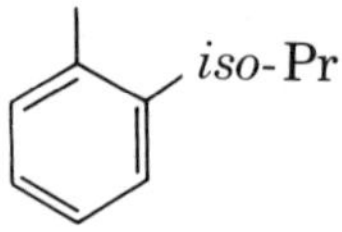 iso-Pr	6
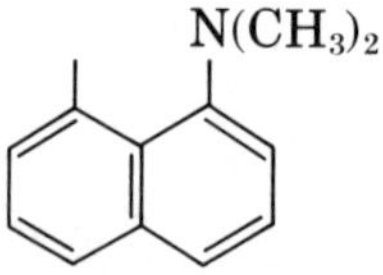 N(CH₃)₂	8	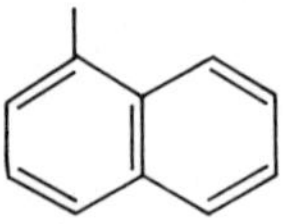	5.5
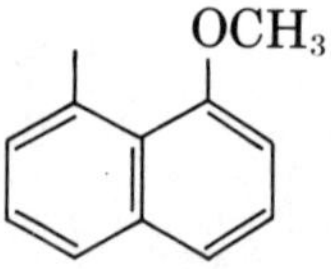 OCH₃	8	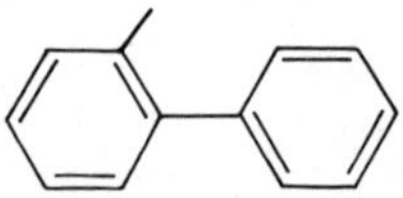	5

[a] For application when the R group relative to a site occupied by the bis-biphenylylene group is in an axial–equatorial orientation in the TP and apical–basal or cis basal–basal orientations in the SP.

[b] In addition, a steric effect between the two bisbiphenylylene groups is applied when the SP transition state (shown for Derivatives **43–52** in Table 1.11) is encountered. This steric term is 4 kcal/mol.

[c] The element effect to the applied for the P–C bonds formed by the biphenylylene ligands is that for the phenyl group (Table 1.7). Also a steric term (Table 1.9) between the unique ligand and the phenyl groups applies.

Table 1.11. Calculated and Observed $\Delta G^{\ddagger}$ for Intramolecular Ligand Exchange for Phosphoranes (70)

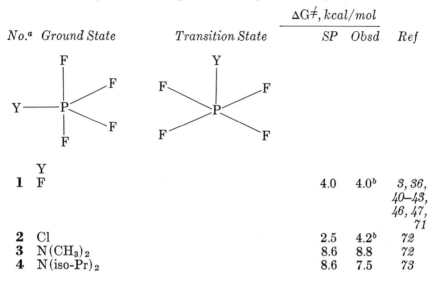

No.[a]	Ground State	Transition State	$\Delta G^{\ddagger}$, kcal/mol SP	Obsd	Ref
	Y				
1	F		4.0	4.0[b]	3, 36, 40–43, 46, 47, 71
2	Cl		2.5	4.2[b]	72
3	N(CH₃)₂		8.6	8.8	72
4	N(iso-Pr)₂		8.6	7.5	73

Transition States

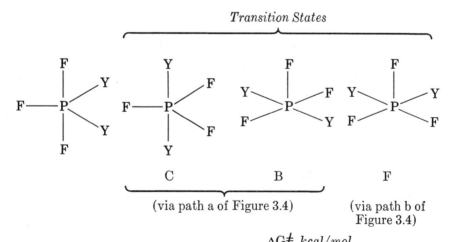

C   B   F

(via path a of Figure 3.4)   (via path b of Figure 3.4)

	Y	$\Delta G^{\ddagger}$, kcal/mol TP C[c]	SP B[c]	SP F[c]	Obsd	
5	Cl	7.2	8.0	8.0	7.2[b]	74
6	H	12.8	11.8	11.8	10.2[b]	75
7	NH₂	19.2	20.0	20.0	—[d]	55
8	N(CH₃)₂	19.2	20.0	26.5	19.6	76
9	Ph	16.4	16.0	17.0	18.7	76
10	CH₃	20.0	18.0	18.0	17.8[b]	76

Table 1.11. Continued

Ground State[e]	*Transition States*	*Ref*

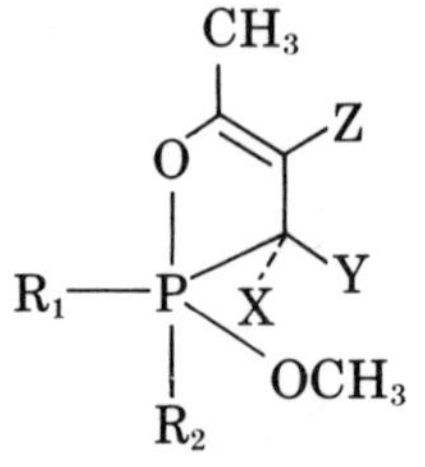

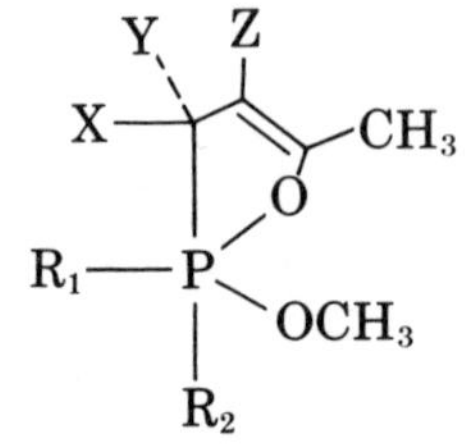

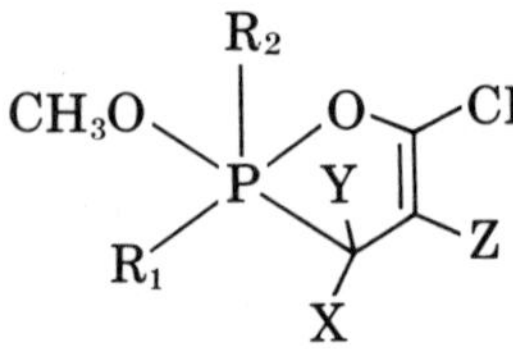

$Z = COCH_3$

	R_1	R_2	X	Y	$\Delta G^{\ddagger}$, kcal/mol		
					TP	*SP*	*Obsd*
11	OCH_3	OCH_3	H	H	8.5	9.1	9.7
12	OCH_3	OCH_3	CH_3	CH_3	12.5	13.1	14.5
13	OCH_3	OCH_3	H	Ph	10.5	11.1	12
14	Ph	OCH_3	H	H	8.5	11.5	9.7
15	Ph	OCH_3	CH_3	CH_3	14.5	17.5	17.2
16	Ph	OCH_3	H	Ph	13.5	16.5	15.8
17	OCH_3	OPh	H	Ph	10.5	11.1	12.5

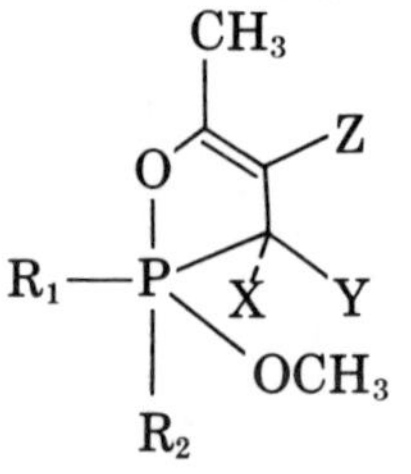

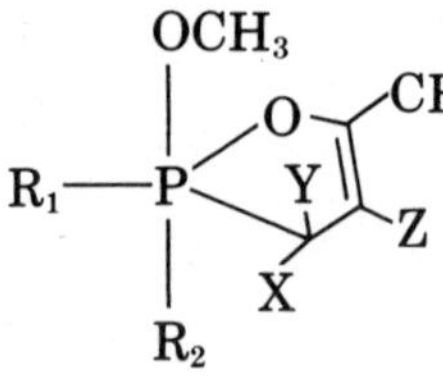

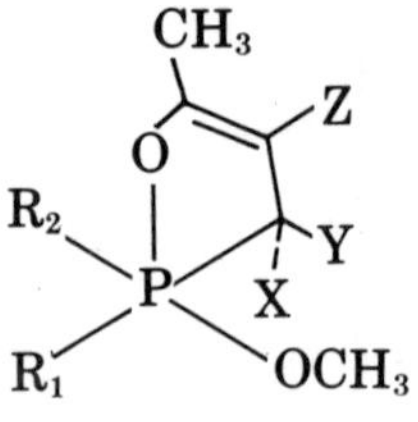

$Z = COCH_3$

	R_1	R_2	X	Y			
18	OCH_3	OPh	H	Ph	21.0	22.1	21.8
19	Ph	OCH_3	H	Ph	21.0	24.5	>24

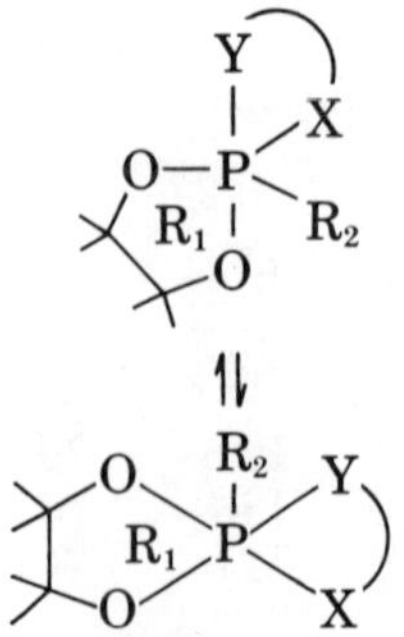

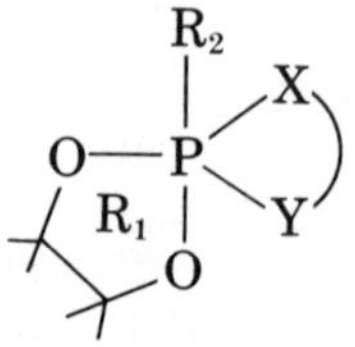

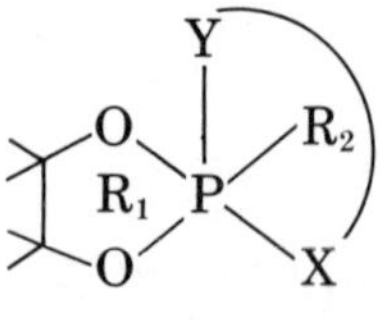

Table 1.11. Continued

	R_1	R_2	$Y \frown X$	$\Delta G\ddagger$, kcal/mol			
				TP	SP	Obsd	Ref
20	CH_3	Ph		8.2	6.7	8.9	
21	CH_3	OPh	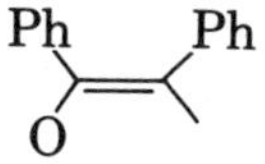	16.0	16.1	14.1	
22	CH_3	OCH_3	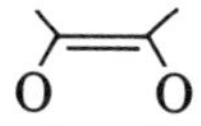	21.0	14.7	21.2	
23′	CH_3	OPh ⎫	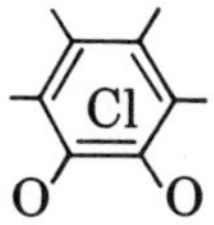	21.0	14.7	20.5	
24	CF_3	OPh ⎭		22.3	14.0	21.8	
25′	CF_3	OPh	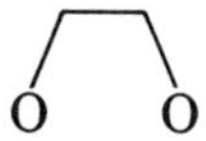	17.3	13.0	17.4	
26	CF_3	$OCH(CF_3)_2$	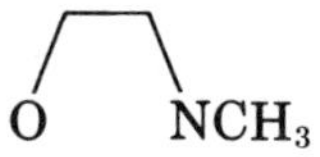 O⏜NCH_3	20.0	18.1	21.0	
27	CF_3	OPh	 $CH_3N⏜NCH_3$	17.4	17.3	20.4	

Ground State[e]   *Transition States*

28

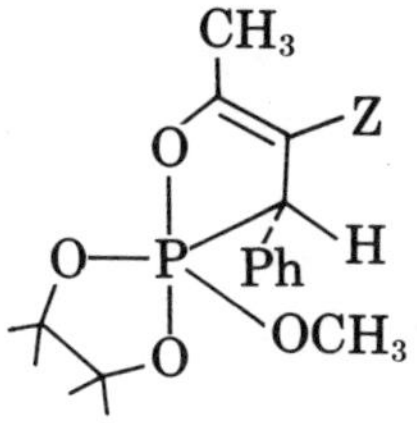

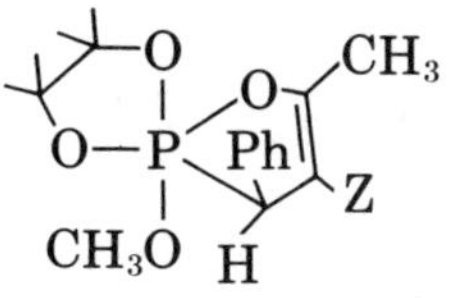

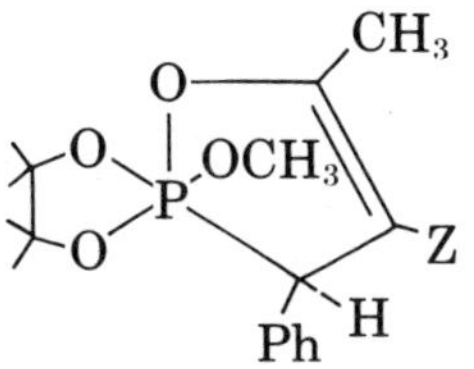

21.0 21.1 17.6 *78, 79*

Table 1.11. Continued

Ground State[e]	Transition States	$\Delta G^{\ddagger}$, kcal/mol			
		TP	SP	Obsd	Ref
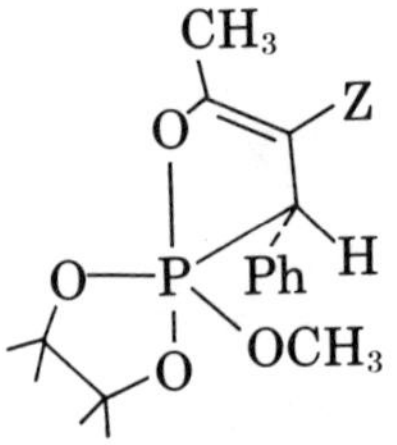 **29**	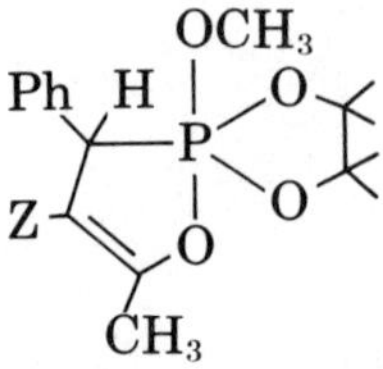 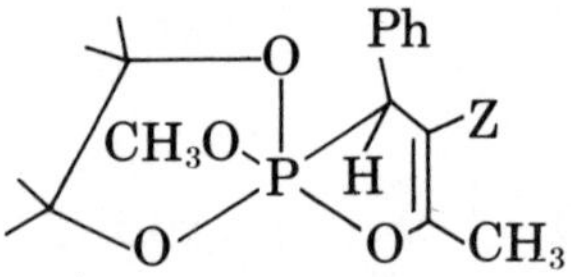	18.0	22.1	>23	78, 79
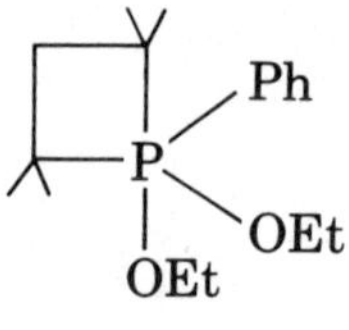 **30**	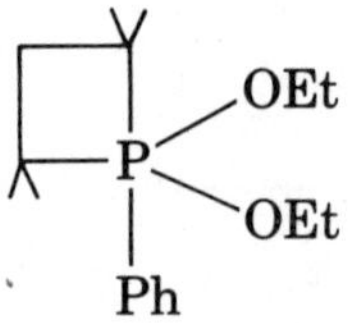 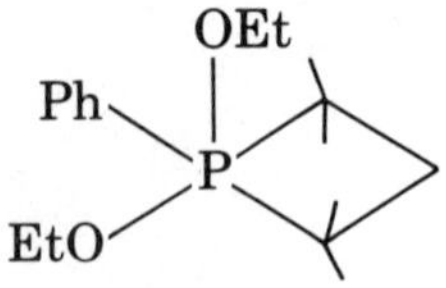	3.7	8.4	11	80, 81

Table 1.11. Continued

Ground State[e]	Transition States	$\Delta G^{\ddagger}$, kcal/mol			
		TP	SP	Obsd	Ref

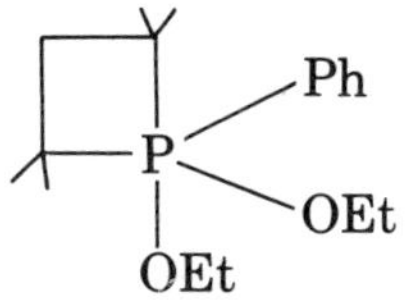

31

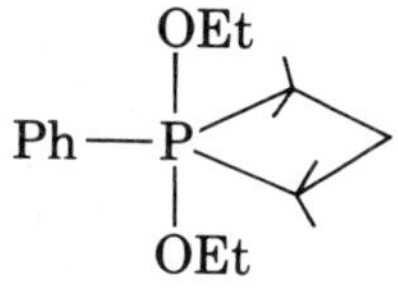

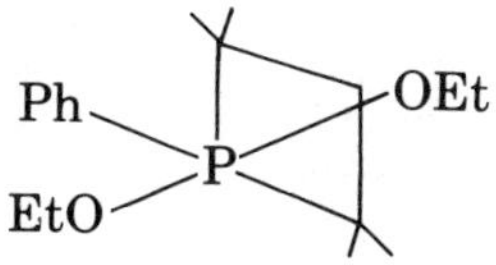

| | | 9.5 | 16.7 | 15 | *82* |

Ground State[e]   *Transition States*

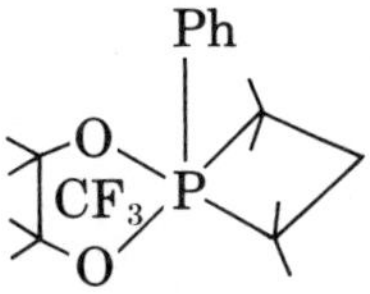

32

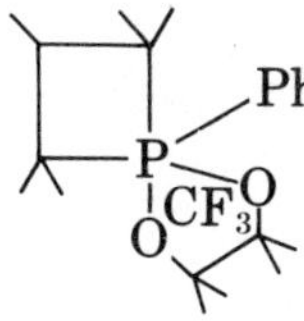

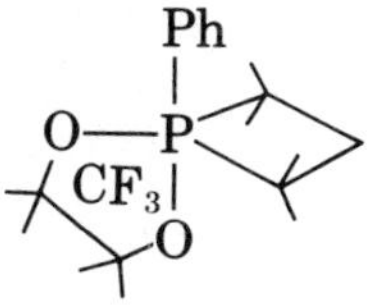

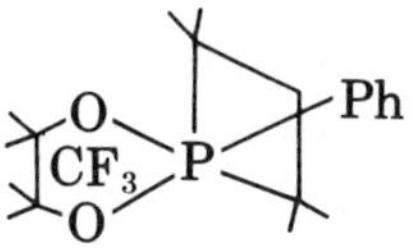

| | | 14.2 | 11.7 | 20.1 | *82* |

| 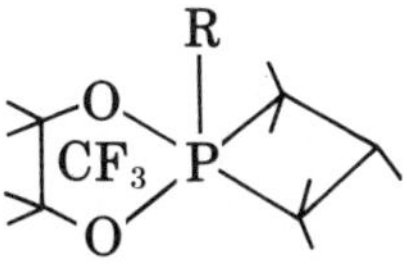 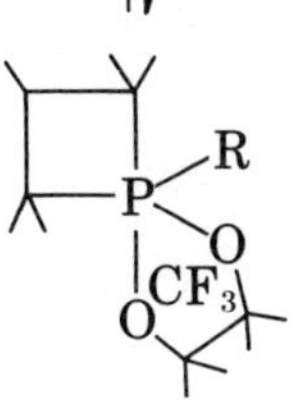 | 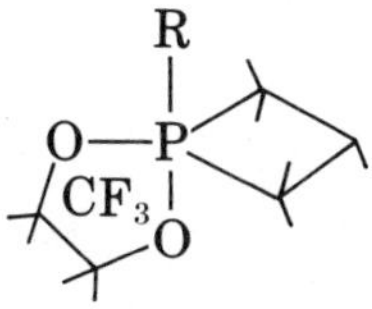 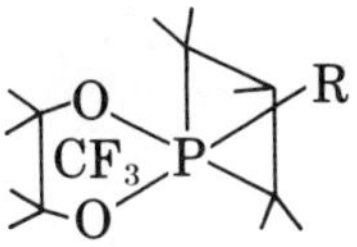 | | | | *83* |

Table 1.11. Continued

$R^{g, h}$	TP	SP	Obsd	Ref
	$\Delta G^{\ddagger}$, kcal/mol			
33 OPh	9.5	6.8	~9	
34 $N(CH_2)_4$	17.6	14.7	16.2	
35 $N(CH_3)_2$	17.6	14.7	16.2	
36 Ph (cis)	14.2	11.7	19.6	
37 Ph	14.2	11.7	>22	
38 $HC{=}C(CH_3)_2$	20.3	14.0	19.1	
39 iso-Pr	20.3	14.0	17.8	
40 CH_3	20.3	14.0	16.9	

Ground State[e]   *Transition States*

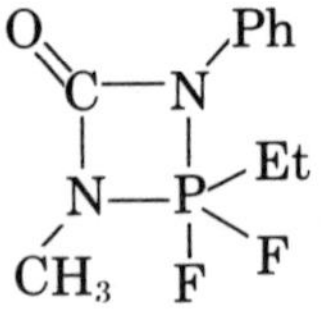

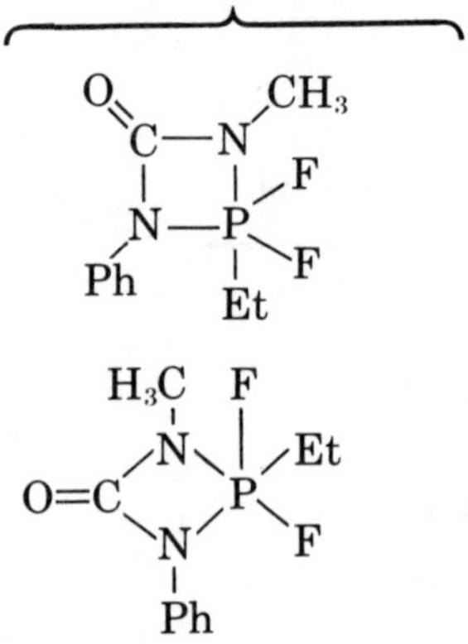

10.0	12.4	11.6	84

Ground State[e]   *Transition States*

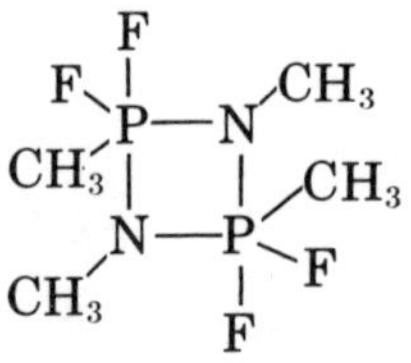

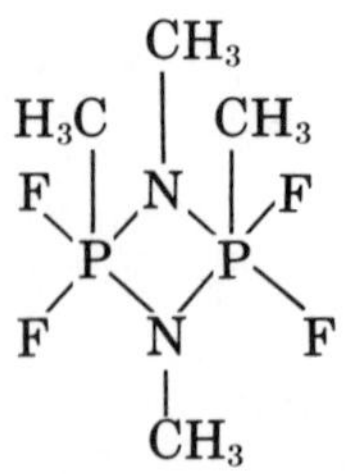

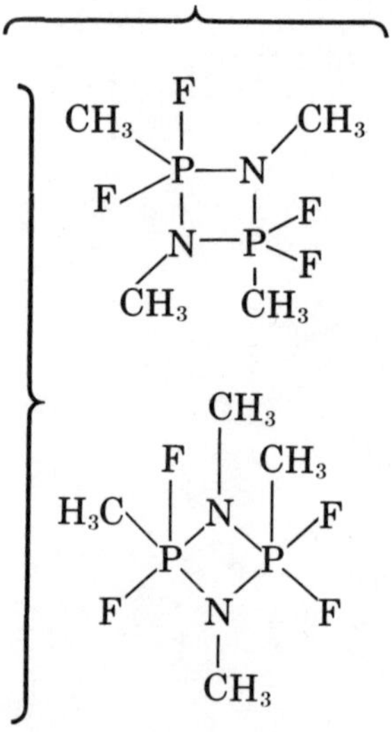

10.0	15.4	~13.3	84

Table 1.11. Continued

Ground State[e]	Transition States	$\Delta G^{\neq}$, kcal/mol		Ref
		SP	Obsd	

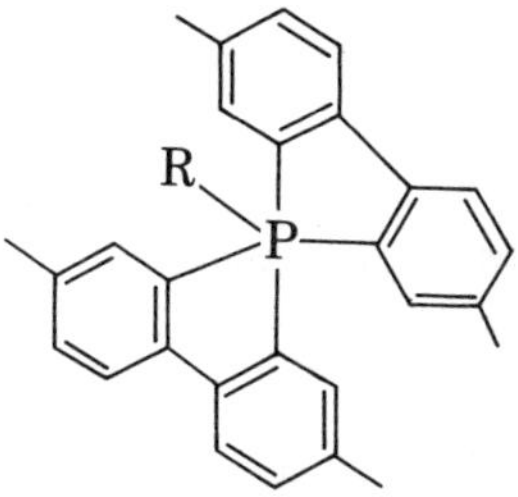

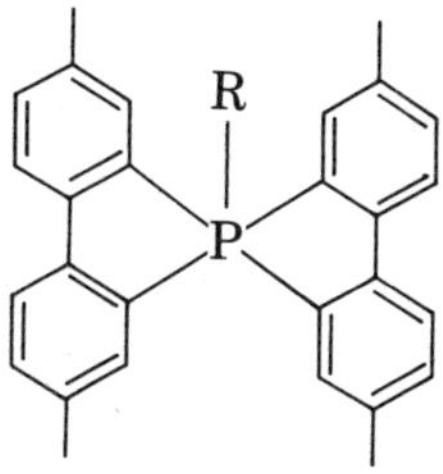

	R			
43	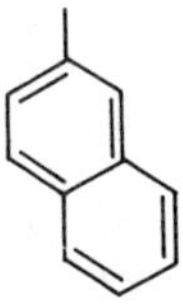	17.9	<11.9	*85*
44	–CH$_3$	12.6	12.5	*85*
45	–C$_2$H$_5$	12.6	13.6	*85*
46	–CH$_2$– 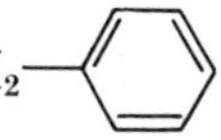	12.6	12.0	*85*
47	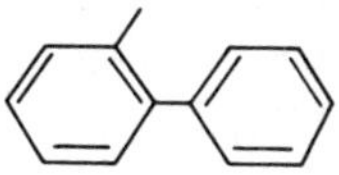	15.9	15.7	*85*
48	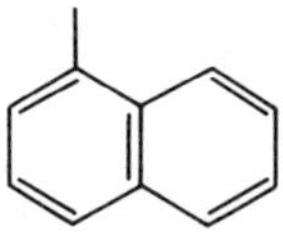	16.9	17.1	*85*
49	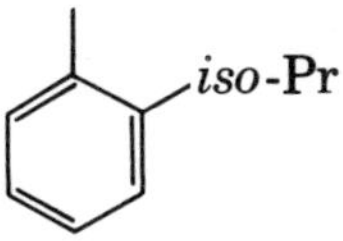 *iso*-Pr	17.9	17.8	*86*
50	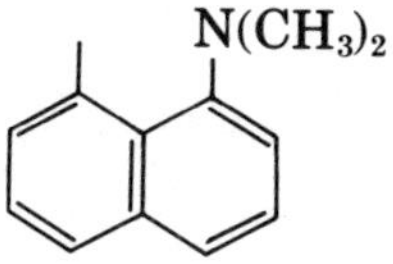 N(CH$_3$)$_2$	21.9	>21	*87*
51	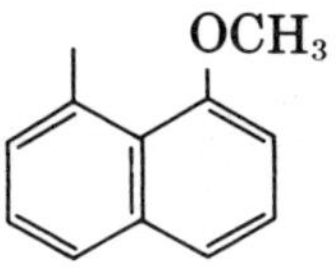 OCH$_3$	21.9	>23	*87*

Table 1.11.   Continued

	$\Delta G^{\neq}$, kcal/mol		
	SP	*Obsd*	*Ref*
52	25.9^i	26.0	*86*

iso-Pr, *iso*-Pr, *iso*-Pr

[a] Bold face italic entries signify that the ground-state SP structure is equal to or lower in energy than the TP. The favored transition state is the one with the highest calculated $\Delta G^{\neq}$ value since both SP and TP structures are traversed in a given path. However, note that two pathways are considered here for Y_2PF_3 derivatives, Entries 5–10.

[b] $\Delta G^{\neq}$ values used in constructing Table 1.7.

[c] The favored transition state is italicized.

[d] Although $\Delta G^{\neq}$ for ligand exchange has not been measured, $\Delta G^{\neq}$ for uncorrelated rotation of the amino groups was determined to be 12.3 kcal/mol (*55*). This value should set a lower limit for the exchange energy.

[e] For these and the following entries in this table only one isomer of each configuration is shown for simplicity—e.g., *cis*- and *trans*-X and Y to R_1—unless stated otherwise. Refer to the respective references for additional detail.

[f] Related derivatives containing $R_2 = SPh$ show values similar to $R_2 = OPh$, implying a similar apicophilicity (*79*). However, this may reflect steric relief in the transition state by the SPh ligand. Further, the ground state of the analogous substance (*88*), $R_1 = CH_3$, $R_2 = NMe_2$, and Y–X =

Ph   Ph / O   O

is considered to be displaced toward the rectangular pyramid (Volume I, Entry **14**, Table 2.3). Additional $\Delta G^{\neq}$ values for other thio derivatives are needed before an element effect, steric terms, and a π bonding contribution can be ascertained.

[g] The x-ray structure of a related substance (Ref *89*) shows a SP conformation. In agreement, the calculation shows an equal or greater stability for the SP relative to the TP ground state for Derivatives **33** and **36–40**.

[h] All derivatives are located trans to 3-CH_3 except OPh, which is reported as a mixture of cis and trans and Ph (cis) as shown.

[i] The SP transition state formulated here for the triisopropyl derivative differs from that postulated in Ref *86*. See Ref *70* for a discussion of this apparently unique example.

Journal of the American Chemical Society

Table 1.12. **Activation Energies for the Epimerization of Chiral Spirophosphoranes**

Ground States *Transition States[a]* *Page*

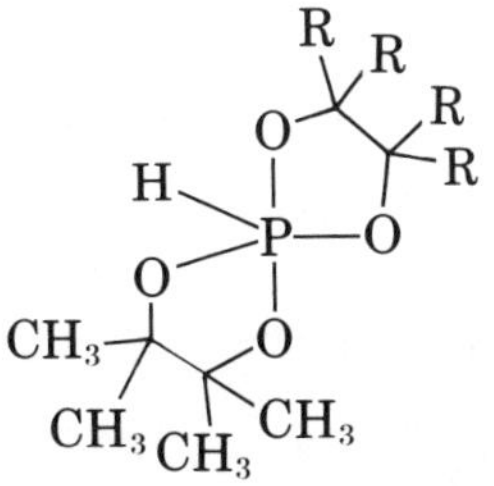

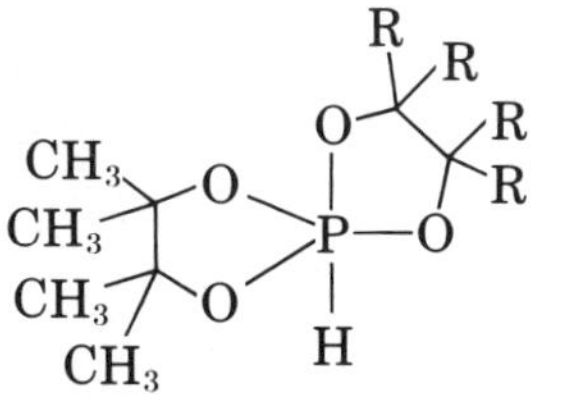

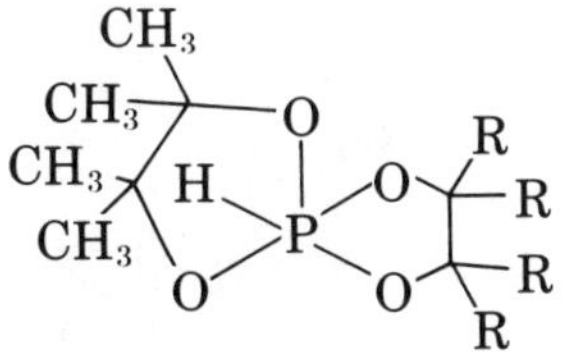

	$\Delta G^{\ddagger}$, kcal/mol			
R	*TP[b]*	*SP[b]*	*Obsd*	
H	21.7	17.6	15.6	Volume I, p 165
CH_3	21.7	17.6	18.4	Volume I, p 165
$1M \rightleftarrows 1P^{c,\,d}$ 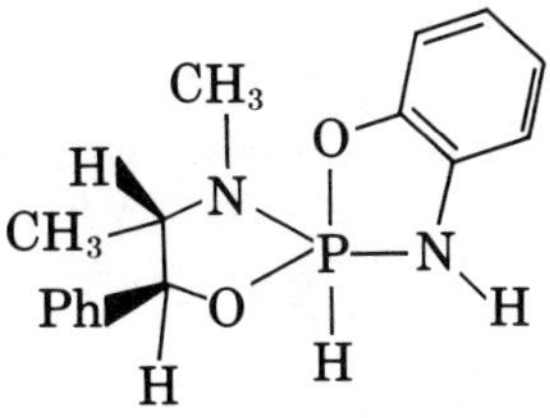 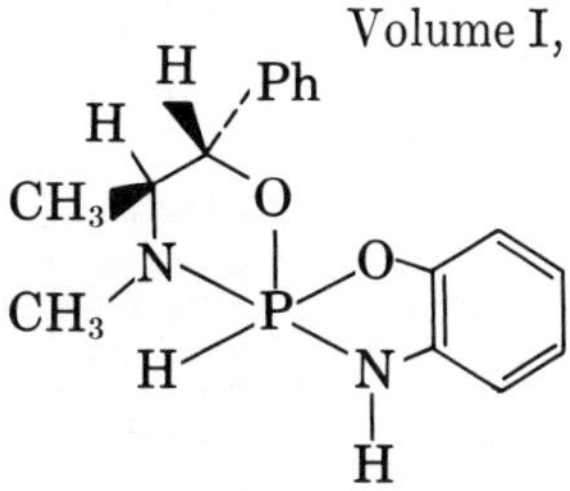				Volume I, p 168
	24.7	25.2	21.5[e]	
			21.2	
$2M \rightleftarrows 2P^c$	24.7	25.2	24.3[e]	Volume I, p 169
			24.2	
$3M \rightleftarrows 3P^f$	24.7	25.2	24.3	Volume I, p 170

Table 1.12.   Continued

	Transition States[a]	Page
Ground States		

	29.9	28.5	28.3
$5P \rightleftarrows 5M$	29.9	28.5	28.3[e] Volume I, p 171
			28.5
$6M \rightleftarrows 6P$	29.9	28.5	27.4[e] Volume I, p 171
			26.7

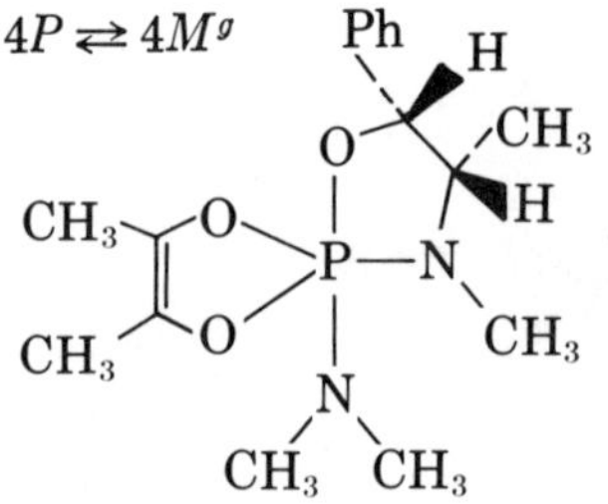

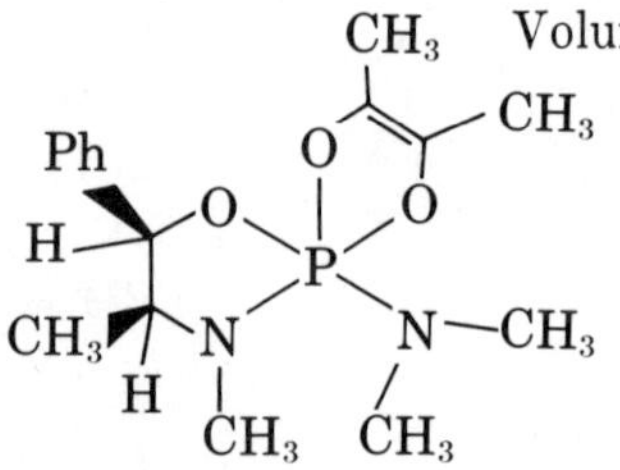

[a] The type of transition state encountered is illustrated. Since there are in general, two comparable epimerization routes for these spirocyclics (cf. Volume I, pp 170 and 172), there will be more than one set of transition state representations. Under the model approximation we are using, these representations will result in the same activation energy.

[b] TP is trigonal bipyramid and SP is square pyramid. The preferred transition state is the one with the highest calculated barrier energy since both of those shown are traversed in the postulated epimerization scheme.

[c] The ground-state configurations are on the pages indicated, with the epimer isolated from solution shown on the left.

[d] Similar transition states represent the epimerization, $2M \rightleftarrows 2P$, and the racemization, $3M \rightleftarrows 3P$.

[e] The upper value of the two $\Delta G^{\ddagger}$ values reported for each substance refers to the forward process, and the lower one refers to the reverse epimerization.

[f] This is a racemization process.

[g] Similar transition states represent the epimerizations, $5P \rightleftarrows 5M$ and $6M \rightleftarrows 6P$.

```
      PROGRAM ISOMER2(INPUT,OUTPUT,TAPE1)
      DIMENSION NAME(30),NATOM(30),ELMAT(12,5),IPI(5),IPIC(5),LATOM(5),IRTYPE(2),JR(2),JR(2),KR(2),ID(50)
     1,RS(10,4),ELMAT(12,5),IPI(5),IPIC(5),LATOM(5),API(3,5),IJC(5),STFAC(39),MSN(25
     15),JSI(15),NN(250),EELEM(50),ERSTRN(50),ERSTRN(50),ETOT(50),ESTERIC(50),INTYPE(
     15),EPI(50),IOK(50),KSI(15),LSI(15),ISTER(5),ISTER(25,25),KSF(25,25),ISI(15)
     1,X(12)
C
C     THIS PROGRAM CALCULATES ISOMER ENERGIES FOR TP AND SP ISOMERS
C     FOR ANY FIVE GIVEN SUBSTITUENTS ON PHOSPHORUS. ISOMER2 CALCULATES
C     THE ENERGIES OF ALL POSSIBLE TP AND SP ISOMERS. ALL CONTRIBUTIONS TO
C     TOTAL ENERGY OF EACH ISOMER ARE TABULATED(I.E.CONTRIBUTION FROM
C     RING STRAIN,SUBSTITUENT EFFECT,STERIC INTERACTIONS,ETC.) A TABLE OF
C     ISOMERS IN ORDER OF DECREASING ENERGY IS PRINTED ON THE REFERENCE
C     J. AM. CHEM. SOC.,100,433(1978)
C
C     THE FIRST DATA CARD HAS THE NAME OF THE MOLECULE(8A10)
C
C     THE SECOND DATA CARD HAS THE SYMBOLS OF SUBSTITUENTS USED IN CALCULATING
C     SUBSTITUENT EFFECT. THE ALLOWED SYMBOLS ARE GIVEN BELOW.
C     AN ASTERISK INDICATES A RING ATOM.
C     CARBON ATOMS
C     CH3
C     C2H5
C     C(CH3)3
C     CH2PH
C     CF3
C     PH
C     C       (USE FOR ALL OTHER CARBON ATOMS)
C
C     OXYGEN ATOMS
C     OH(H+)
C     OCH3
C     OCH3(H+)
C     OC2H5
C     OC2H5(H+)
C     OPH(H+)
C     O(I-PR)          (I-PR = ISO-PROPYL)
C     O(T-BU)          (T-BU) = TERT-BUTYL
C     OC*(H+)
C     OSICL3
C     O-      (USE FOR ALL OTHER OXYGEN ATOMS)
```

```
C     OTHER ATOMS
C         H
C         CL
C         N
C         SH
C         SICL3
C         LP    (LONE PAIR)
C         ELECTRON  (ODD ELECTRON)
C
C     THE FORMAT IS(4X,5(A10,5X))   PUT THE FIRST CHARACTER OF THE SYMBOL IN
C     THE FIRST BLANK OF THE A FIELD. IT IS IN COLUMN 5,20,35,50,65. IN WHICH
C     THE POSITION NUMBER OF THE GROUP IS DETERMINED BY THE ORDER IN WHICH
C     THE GROUPS ARE NAMED.
C
C     THE THIRD DATA CARD TELLS
C        NUMBER OF RINGS IN COMPOUND,NR
C        NUMBER OF ISOMERS TO BE CALCULATED,N
C        NUMBER OF PI BONDING EFFECTS,M
C        STERIC INTERACTION OPTION,NSE
C        OPTION TO INCLUDE FOOTNOTE(A),MFU
C     FORMAT (5I5)
C
C     NR TELLS HOW MANY RINGS ARE IN THE COMPOUND
C
C     N TELLS HOW MANY ISOMERS ARE TO BE CALCULATED.  SET N=0 TO CALCULATE
C     ALL ISOMERS.
C
C     M SPECIFIES THE NUMBER OF LIGANDS FOR WHICH PI BONDING EFFECTS ARE
C     TO BE INCLUDED
C
C     NSE DETERMINES WHICH OPTION TO USE FOR STERIC INTERACTIONS.
C     IF NSE=0,NO STERIC INTERACTIONS
C     IF NSE=1,SPECIFIC INTERACTIONS ARE ENTERED
C     IF NSE=2,ATTACHED GROUPS ARE ENTERED BY IDENTIFYING NUMBER AND
C     ALL STERIC, INTERACTIONS ARE CALCULATED.
C
C     MFU TELLS IF CERTAIN STERIC INTERACTIONS SHOULD BE MULTIPLIED BY 0.75.
C     IT IS USED FOR THE SPECIAL CASE WHEN THERE ARE TWO IDENTICAL INTERACTIONS
C     INVOLVING AN APICAL/AXIAL ATOM(NOT A MEMBER OF A RING) AND THIS APICAL/AXIAL
C     ATOM HAS NO OTHER STERIC INTERACTIONS. SEE FOOTNOTE (A) TABLE V.
C     SET MFU = 0 IF YOU DO NOT WANT TO INCLUDE THIS FACTOR.
C     IF MFU = 1 OR MORE, THE FACTOR WILL BE APPLIED EVERY TIME THE STERIC
```

```
C     INTERACTION MEETS ALL THE SPECIAL CONDITIONS DESCRIBED ABOVE.
C
C     THE FOURTH SET OF DATA CARDS IDENTIFIES RINGS.
C     IF NR = 0, THE FOURTH SET OF DATA CARDS IS NOT USED.
C     THESE CARDS IDENTIFY RING TYPE AND GIVE THE POSITION NUMBERS OF THE TWO
C     POINTS OF ATTACHMENT. THERE IS A SEPARATE CARD FOR EACH RING.
C     FORMAT(3I5)
C     THE RING TYPE IS IDENTIFIED BY NUMBER--EACH RING NUMBER (1-9) CORRESPONDING
C     TO AN ENTRY IN TABLE 5.8. RING TYPE 10 IS AN UNSATURATED 5-MEMBERED
C     RING WITH O ATTACHED TO P.
C
C     THE FIFTH SET OF DATA CARDS IDENTIFIES THE ISOMERS TO BE CALCULATED.
C     IF N=0, THIS SET OF DATA CARDS IS NOT USED.
C     THESE CARDS IDENTIFY THE ISOMERS SUCH AS TP12,TP21,SP12345, ETC.
C     FORMAT(I5)
C     SP ISOMERS ARE IDENTIFIED BY GIVING APICAL ATOM FIRST FOLLOWED BY
C     BASAL ATOMS IN COUNTERCLOCKWISE DIRECTION. TP ISOMERS ARE IDENTIFIED BY
C     GIVING AXIAL ATOMS FIRST. AS YOU LOOK FROM THE FIRST AXIAL ATOM
C     TO THE SECOND, IT IS ASSUMED THAT THE COUNTERCLOCKWISE DIRECTION IS THE
C     DIRECTION OF INCREASING ATOM NUMBERS.
C
C     THE SIXTH SET OF DATA CARDS IDENTIFIES PI EFFECTS.
C     IF M=0, THE SIXTH SET OF CARDS IS NOT USED.
C     THE M POSITION IS NOT EQUAL TO ZERO. IDENTIFY THE PI LIGAND BY ITS TYPE AND IDENTIFY
C     THE PI LIGAND BY THE POSITION NUMBER OF THE LIGAND.
C     TYPE 1 = NR2,NH2
C     TYPE 2 = RN+P
C     TYPE 3 = SR
C     FORMAT (15I5)
C
C     THE SEVENTH SET OF DATA CARDS IS USED FOR STERIC FACTORS.
C     IF NSE=0, NO STERIC INTERACTIONS ARE INCLUDED AND THESE CARDS ARE NOT USED.
C     IF NSE=1, OPTION 1 IS USED.
C     IF NSE=2, OPTION 2 IS USED.
C     FOR OPTION 1, READ MS= (I5 FORMAT)
C     MS IS THE TOTAL NUMBER OF STERIC INTERACTIONS.
C     IF MS = 0, NO MORE CARDS ARE NEEDED IN THIS SET.
C     THE REST OF THE CARDS DESCRIBE THE STERIC INTERACTION BY SPECIFIC NUMBER
C     AND TELL WHICH TWO ATOMS ARE INVOLVED.
C     SEE THE LISTING OF STERIC FACTORS (1-39) GIVEN BELOW FOR NUMBER CODE TO
C     IDENTIFY SPECIFIC STERIC INTERACTIONS.
C     FORMAT (15I5)
```

```
C  IF THE INTERACTION IS BETWEEN 2 BIPHENYLENE GROUPS (TYPE 31) IDENTIFY
C  THE NUMBER OF THE ATOM THAT IS NOT BONDED TO BIS-BIPHENYLENE
C  AND THEN PUT A ZERO IN THE NEXT POSITION.
C
C  FOR OPTION 2 READ ISTER(I),I=1,5  (5I5 FORMAT)
C  ISTER(I) IS THE NUMBER OF THE GROUP ATTACHED TO ATOM (I)  USE THE FOLLOWING
C  NUMBERS TO IDENTIFY THE GROUPS
C  * INDICATES A RING ATOM
C
C  11  - CH3
C  16  - C2H5
C  24  - C(CH3)3
C  25  - CH2(PH)
C  43  - C*(CH3)2
C   7  - G*(H)(PH)
C   1  - PH   PH = PHENYL
C  17  - BOND TO BIS BIPHENYLENE
C  18  - SEE ENTRY 1, TABLE  5.10
C  19  - SEE ENTRY 2, TABLE  5.10
C  20  - SEE ENTRY 3, TABLE  5.10
C  21  - SEE ENTRY 4, TABLE  5.10
C  22  - SEE ENTRY 5, TABLE  5.10
C  23  - SEE ENTRY 6, TABLE  5.10
C   8  - OR(PH)    R = CH3, C2H5
C  10  - O-C(=H)(CF3)2
C  14  - X=C(PH)=C(PH)-X-    (X=0,C)
C  15  - X=C(CF3)2-C(CF3)2-X-  (X=0,C)
C  16  - NR2  (R=CH3,C2H5)
C   5  - N+-PH
C   9  - N+-CH3
C  13  - F
C  12  - CL
C   0  - (ZERO) INDICATES ALL OTHER GROUPS
C
C  DATA FOR MORE THAN ONE MOLECULE CAN BE CALCULATED.  AFTER THE LAST
C  MOLECULE, PUT A DATA CARD WITH XX IN COL 1-2
C
C  SET UP ARRAY OF ELECTRONEGATIVITY VALUES
C  DATA X/1.5,2.0,2.5,2.6,2.7,2.8,3.0,3.1,3.5,3.7,3.8,4.0/
C
C  SET UP THE TABLE FOR ESTIMATING RING STRAIN VALUES.  THERE ARE TEN ROWS
C  IN THE TABLE, CORRESPONDING TO THE NINE ENTRIES IN TABLE 5.8 AND TO A
```

```
C     TENTH ENTRY FOR A SATURATED C10 HETEROCYCLE.
C     THE FOUR COLUMNS GIVE INTERACTIONS FOR AX-EQ,EQ-EQ,CIS BAS-BAS, AND
C     AP-BAS RESPECTIVELY. VALUES ARE READ IN A COLUMN AT A TIME;
C     THAT IS, THE FIRST NINE VALUES ENTERED HERE ARE FROM COLUMN 1 OF TABLE 5.8.

      DATA RS/8.,9*2.,
     124.,0.,15.,18.,21.,24.,15.,21.,23.,16.5,
     12.,0.,9*1.0,
     116.,0.,0.,9.0,14.,14.,9.,13.,15.,11.5/

C     SET UP THE TABLE FOR ESTIMATING SUBSTITUENT EFFECT.   THERE ARE TWELVE
C     ROWS IN THE TABLE, CORRESPONDING TO TWELVE TYPES OF SUBSTITUENT GROUPS.
C     THERE ARE 5 COLUMNS IN THE TABLE (TPEQ,TPAX,SPAP,SPBAS,SPBAS).
C     COLUMNS 4 AND 5 ARE IDENTICAL. VALUES ARE READ IN A COLUMN AT A TIME, I.E.
C     THE FIRST TWELVE ENTRIES ARE VALUES FROM COLUMN 1 ( VALUES FOR TPEQ).

      DATA ELMAT/-1.,5.,0.,0.,1.,3.,1.,5.,1.,7.,1.,9.,2.,2.,4.,2.,8.,3.,0.,3.,0.,3.,0.,
     114.,5.,7.,4.,7.,3.,2.,6.,3.,3.,0.,3.,8.,1.,0.,5.,0.,
     1-2.,4.,0.,2.,3.,2.,6.,1.,9.,1.,7.,1.,5.,1.,2.,1.,0.,0.,4.,0.,2.,0.,1.,0.,0.,
     117.,4.,4.,0.,2.,2.,1.,9.,1.,7.,1.,5.,1.,2.,1.,0.,0.,4.,0.,2.,0.,1.,0.,0./

C     SET UP TABLE FOR PI EFFECTS(3 X 5 MATRIX)  THE THREE ROWS ARE FOR LIGANDS
C     NR2/NH2,RN*P,SCH3. THE COLUMNS TELL THE AMOUNT OF ENERGY DUE TO LOSS OF PI
C     BONDING FOR LOCATIONS IN TP AX, SP BAS(TRANS TO TYPE 4), THE COLUMNS ARE FOR LOCATIONS
C     TP EQ, SP AX, SP BAS(TRANS TO TYPE 5),  SP BAS(TRANS TO TYPE 5)
C     VALUES ARE ENTERED IN ONE COLUMN AT A TIME.

      DATA API/3*0.0,4*(6.0,2.0,2.0)/

C     SET UP TABLE FOR STERIC FACTORS.
C     THESE VALUES ARE USED FOR OPTION 1 AND 2.       THE STERIC INTERAC-
C     SEE TABLE 9 AND 10 FOR EXPLANATION OF FACTORS.  ASTERISK INDICATES-
C     TIONS ARE IDENTIFIED BY THE FOLLOWING NUMBERS (    ASTERISK INDICATES
C     A RING ATOM)
C     THE VALUES OF THE STERIC FACTORS ARE ALSO LISTED HERE.

      1    R2N-P-NR2                        6.5
      2    OMIT--SEE INTERACTION 13
      3    PH-P-PH              13          6.0
      4    PH-P-C*-(CH3)2                   6.0
      5    PH-P-C*-H(PH)                    5.0
      6    PH-N*-P-(C2H5)                   5.0
      7    R-P-PH                           4.0
```

```
 8   RO-P-C*(CH3)2                               4.0
 9   (CH3)N*-P-PH                                 4.0
10   (CH3O)-P-C*-(CH3)2                           3.0
11   (CH3)N*-P-(C2H5)                             3.0
12   (CH3)N*-P-OPH                                3.0
13   (CH3)2-N-P-C*-(CH3)2                         4.0
14   (CH3)2N*-P-(CH3)                             2.0
15   (PH+O)-P-C*-H(PH)                            2.0
16   RO-P-C*-H(PH)                                2.0
17   (CH3)2N*-P-CL                                2.0
18   F-P-PH                                       2.0
19   RO-P-PH                                      1.0
20   RO-P-OR                                      1.0
21   RO-P-(O-PH)                                  1.0
22   (CH3)2N*-P-O-C-(H)(CF3)2                     1.0
23   SEE ENTRY 23--TABLE 5.9                      3.0
24   SEE ENTRY 24--TABLE 5.9                      2.0
25   SEE ENTRY 1--TABLE 5.10                     10.0
26   SEE ENTRY 3--TABLE 5.10                      8.0
27   SEE ENTRY 3--TABLE 5.10                      8.0
28   SEE ENTRY 4--TABLE 5.10                      6.0
29   SEE ENTRY 5--TABLE 5.10                      6.0
30   SEE ENTRY 6--TABLE 5.10                      5.0
31   BIPHEN-BIPHEN-(FOOTNOTE (B) TABLE 5.10       4.0
32   (CH3)3-C-P-C(PH)3                            8.0
33   (CH3)3-C-P-(PH2-PH)                          7.0
34   (CH3)3-C-P-(CH2-PH)                          6.0
35   (PH-)-P-(CH2-PH)                             5.0
36   (PH-)-P-(CH2-CH3)                            5.0
37   R2N*-P-CH3                                   2.0
38   (PH-)(HOR)-P-CH3                             2.0
39   (PH-CH2)-P-C*(CH3)2                          3.0

      DATA (STFAC(I),I=1,39)/6.5,0.,2*6.0,2*5.0,2*6.0,2*5.0,0.,3*4.0,3*3.0,4.0,5,*2.0,
     114*1.0,3.0,2.0,10.,2*8.0,6.0,5.5,5.0,4.0,8.0,7.0,6.0,2*3.0,0.,0.,
     112.0,3.0/

C     SET UP TABLE TO IDENTIFY THE NUMBER OF THE STERIC INTERACTION
C     THIS TABLE IS USED IN OPTION 2. THE NUMBERS ARE STORED IN KSF(I,J)
C     AND IDENTIFY THE GROUPS FOR WHICH STERIC EFFECT IS TO BE INCLUDED. THE
C     FIRST 25 VALUES ARE COLUMN 1 OF THE ARRAY. ALL VALUES ARE STORED WITH J
C     GREATER THAN OR EQUAL TO I.
```

```
      DATA KSF/1,24*0,13,24*0,0,4,3,22*0,
     12*0,5,22*0,25*0,2*0,7,0,6,20*0,
     125*0,38,8,19,16,3*0,20,17*0,2*0,9,2*0,11,19*0,
     10,10,0,15,3*0,21,12,16*0,2*0,7,5*0,14,16*0,8*0,17,16*0,
     12*0,18,22*0,8*0,22,16*0,14*0,23,10*0,
     115*0,24,9*0,0,4,3,5,0,7,0,19,9,0,7,0,18,3*0,31,8*0,16*0,25,8*0,
     116*0,26,8*0,16*0,27,8*0,16*0,28,6*0,
     116*0,29,8*0,16*0,30,8*0,2*0,33,7*0,35,5*0,33,6*0,32,0,
     10,39,36,7*0,37,5*0,36,6*0,34,0/
C
C     SET UP THE ISOMER TABLE
C
      DATAID/4HTP12,4HTP21,4HTP13,4HTP31,4HTP14,
     14HTP41,4HTP15,4HTP51,4HTP23,4HTP32,
     14HTP24,4HTP42,4HTP25,4HTP52,4HTP34,
     14HTP43,4HTP35,4HTP53,4HTP45,4HTP54,
     17HSP12345,7HSP12543,7HSP12435,7HSP12534,7HSP12354,7HSP12453,
     17HSP21345,7HSP21543,7HSP21435,7HSP21534,7HSP21354,7HSP21453,
     17HSP31425,7HSP31524,7HSP31245,7HSP31542,7HSP31254,7HSP31452,
     17HSP41325,7HSP41523,7HSP41235,7HSP41532,7HSP41253,7HSP41352,
     17HSP51324,7HSP51423,7HSP51234,7HSP51432,7HSP51243,7HSP51342/
C
C     SET UP THE TABLE FOR NTYPE VALUES FOR EACH ISOMER
C
C
C     THE NN MATRIX IDENTIFIES THE POSITIONS IN EACH OF THE ISOMERS
C     2 = TP AX
C     1 = TP EQ
C     3 = SP AP
C     4 = SP BAS ( TRANS TO ANOTHER 4)
C     5 = SP BAS (TRANS TO ANOTHER 5)
C     THE FIRST FIVE ENTRIES ARE FOR TP12,NEXT FIVE FOR TP21, ETC.
C     ISOMERS ARE IN THE SAME ORDER AS IN ISOMER TABLE ABOVE.
      DATANN/2,2,1,1,1,2,2,1,1,1,2,1,2,1,1,2,1,2,1,1,2,1,1,2,1,
     12,1,1,2,1,2,1,1,1,2,2,1,1,1,2,1,2,2,1,1,1,2,2,1,1,
     11,2,1,2,1,1,2,1,2,1,1,2,1,1,2,1,2,1,1,2,1,1,2,2,1,
     11,1,2,2,1,1,1,2,1,2,1,1,2,1,2,1,1,1,2,2,1,1,1,2,2,
     13,4,5,4,5,3,4,5,4,5,3,4,4,5,5,3,4,4,5,5,3,4,5,5,4,3,4,5,5,4,
     14,3,5,4,5,4,3,5,4,5,4,3,4,5,5,4,3,4,5,5,4,3,5,5,4,4,3,5,5,4,
     14,4,3,5,5,4,4,3,5,5,4,5,3,4,5,4,5,3,4,5,4,5,3,5,4,4,5,3,5,4,
     14,4,5,3,5,4,4,5,3,5,4,5,4,3,5,4,5,4,3,5,4,5,5,3,4,4,5,5,3,4,
     14,4,5,5,3,4,4,5,5,3,4,5,4,5,3,4,5,4,5,3,4,5,5,4,3,4,5,5,4,3/
C
C
C      PRINT SUBSTITUENT EFFECT TABLE
```

```
C
      PRINT 600
      PRINT 4000
4000  FORMAT(/*      TABLE OF SUBSTITUENT EFFECTS*/)
      PRINT 4001
4001  FORMAT(*  SUBSTITUENT TYPE   ELECTRONEGATIVITY    TP(EQ)    TP(AX)
     1          SP(AP)
     1  SP(BAS)*)
      DO 4003 I = 1,12
4003  PRINT 4002,I,X(I),(ELMAT(I,J),J=1,5)
4002  FORMAT(10X,I2,8X,1X,F10.1,4X,5(F10.1))
C
C     PRINT FI EFFECT TABLE
C
      PRINT 4010
4010  FORMAT(//*        TABLE OF PI EFFECTS*)
      PRINT 4011
4011  FORMAT(*  LIGAND TYPE  SP(BAS)  TP(EQ)      TP(AX)
     1  SP(BAS)*)
      DO 4012 I=1,3
4012  PRINT 4013,I,(API(I,J),J=1,5)
4013  FORMAT(6X,I2,7X,5(F8.1,7X))
C
C     PRINT TABLE OF STERIC INTERACTION VALUES
C
      PRINT 4020
4020  FORMAT(//* TABLE OF VALUES FOR STERIC INTERACTIONS*)
      PRINT 4030
4030  FORMAT(*  APPLIED FOR TP(AX-EQ), SP(AP-BAS), SP(CIS BAS-BAS)
     1 INTERACTIONS*)
      PRINT 4021
4021  FORMAT(*   TYPE OF INTERACTION     FACTOR*)
      DO 4022 I=1,39
4022  PRINT 4023,I,STFAC(I)
4023  FORMAT(8X,I4,8X,F7.2)
C
C     PRINT RING STRAIN TABLE
C
      PRINT 4005
4005  FORMAT(//*     RING STRAIN VALUES*)
      PRINT 4006
4006  FORMAT(*  RING TYPE    TP(AX-EQ)    TP(EQ-EQ)    SP(CIS BAS-B
     1AS)  SP(AP-BAS)*)
      DO 4007 I=1,10
4007  PRINT 4008,I,(RS(I,J),J=1,4)
4008  FORMAT(6X,I2,7X,4(F8.1,7X))
```

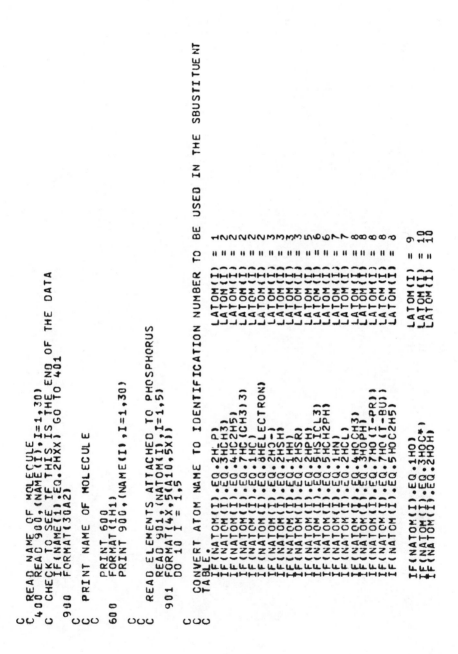

```
C   READ NAME OF MOLECULE
400 READ 900,(NAME(I),I=1,30)
C   CHECK TO SEE IF THIS IS THE END OF THE DATA
    IF(NAME(1).EQ.2HXX) GO TO 401
900 FORMAT(30A2)
C
C   PRINT NAME OF MOLECULE
C
    PRINT 600
600 FORMAT(1H1)
    PRINT 900,(NAME(I),I=1,30)
C
C   READ ELEMENTS ATTACHED TO PHOSPHORUS
    READ 901,(NATOM(I),I=1,5)
901 FORMAT(4X,5(A10,5X))
    DO 10 I = 1,5
C
C   CONVERT ATOM NAME TO IDENTIFICATION NUMBER TO BE USED IN THE SBUSTITUENT
    TABLE.
    IF(NATOM(I).EQ.2H P)      LATOM(I) = 1
    IF(NATOM(I).EQ.3HCH3)     LATOM(I) = 2
    IF(NATOM(I).EQ.4HC2H5)    LATOM(I) = 2
    IF(NATOM(I).EQ.7HC(CH3)3) LATOM(I) = 2
    IF(NATOM(I).EQ.1HC)       LATOM(I) = 2
    IF(NATOM(I).EQ.8HELECTRON)LATOM(I) = 3
    IF(NATOM(I).EQ.2HO-)      LATOM(I) = 3
    IF(NATOM(I).EQ.2HSH)      LATOM(I) = 3
    IF(NATOM(I).EQ.1HH)       LATOM(I) = 3
    IF(NATOM(I).EQ.2HSR)      LATOM(I) = 5
    IF(NATOM(I).EQ.2HPH)      LATOM(I) = 6
    IF(NATOM(I).EQ.5HSICL3)   LATOM(I) = 7
    IF(NATOM(I).EQ.5HCH2PH)   LATOM(I) = 7
    IF(NATOM(I).EQ.1HN)       LATOM(I) = 8
    IF(NATOM(I).EQ.2HCL)      LATOM(I) = 8
    IF(NATOM(I).EQ.4HOCH3)    LATOM(I) = 8
    IF(NATOM(I).EQ.3HOPH)     LATOM(I) = 8
    IF(NATOM(I).EQ.7HO(I-PR)) LATOM(I) = 8
    IF(NATOM(I).EQ.7HO(T-BU)) LATOM(I) = 8
    IF(NATOM(I).EQ.5HOC2H5)   LATOM(I) = 8
    IF(NATOM(I).EQ.1HO)       LATOM(I) = 9
    IF(NATOM(I).EQ.3HOC*)     LATOM(I) = 10
    IF(NATOM(I).EQ.2HOH)      LATOM(I) = 10
```

```
      IF(NATOM(I).EQ.8HOCH3(H,))    LATOM(I) = 10
     1IFNATOM(I).EQ.9HOC2H5(H,))    LATOM(I) = 11
     1IFNATOM(I).EQ.7HOC(H,))       LATOM(I) = 11
     1IFNATOM(I).EQ.6HOCH(CL,3)     LATOM(I) = 12
     1IFNATOM(I).EQ.3HCF3)          LATOM(I) = 12
     1IFNATOM(I).EQ.1HF)            LATOM(I) = 12
   10 CONTINUE
      PRINT 902
      DO 11 I=1,5
  902 FORMAT(903,(NATOM(I),LATOM(I))
  903 FORMAT(3X,I2,10X,A10,5X,I5)    SUBSTITUENT TYPE*)
C
C     INPUT FOR NUMBER OF RINGS,NUMBER OF ISOMERS,NUMBER OF PI BONDING
C     EFFECTS,STERIC INTERACTION OPTION,FOOTNOTE(A) OPTION (NR,N,M,NSE,MFU)
      READ 904,NR,N,M,NSE,MFU
C
C     INPUT FOR RINGS
      IF(NR.EQ.0) GO TO 16
      DO 12 I=1,NR
   12 READ 904,KRTYPE(I),JR(I),KR(I)
C
C     INPUT FOR NUMBER OF ISOMERS
   16 CONTINUE
  904 FORMAT(1515)
      IF(N.EQ.0) GO TO 13
C
C     KEEP ID VALUES FOR LATER USE WITH OTHER MOLECULES
      DO 436 I=1,N
  436 IDK(I) = ID(I)
      READ 205,(ID(I),I=1,N)
  905 FORMAT(8A10)
      GO TO 14
   13 N = 50
   14 CONTINUE
C
C     INPUT FOR PI INTERACTIONS
      DO 815 I=1,5
      IPI(I) = 0
      IJ(I)=0
  815 CONTINUE
      IF(M.EQ.0) GO TO 15
      READ 904,(IPI(I),IJ(I),I=1,M)
   15 CONTINUE
C
C     INPUT FOR STERIC EFFECTS
C     IF STERIC EFFECTS ARE TO BE ENTERED BY SPECIFIC INTERACTION NUMBER,
C     NSE=1.
C     IF THE GROUP ON EACH ATOM IS TO BE READ IN, AND NUMBER AND TYPE OF
C     STERIC INTERACTION CALCULATED BY PROGRAM,NSE=2.
```

```
C     IF NO STERIC INTERACTIONS ARE BEING CONSIDERED NSE=0.
      IF(NSE-1)2002,2,00,2001
2002  GO TO 17
C
C     OPTION 1 FOR STERIC INTERACTIONS
C
2000  READ 904,MS
      IF(MS.EQ.0) GO TO 17
      READ 904,(MSN(I),ISI(I),JST(I),I = 1,MS)
      GO TO 17
C
C     OPTION 2 FOR STERIC INTERACTIONS
C
2001  MS = 0
      READ 904,(ISTER(I),I=1,5)
      NSTOP = 0
      DO 5000 I = 1,5
      IST = ISTER(I)
C
C     ELIMINATE INTERACTION IF GROUP HAS NOT BEEN SPECIFIED
      IF(IST.EQ.0) GO TO 5000
      DO 6000 J = 1,5
      IF(I.EQ.J) GO TO 6000
      JST = ISTER(J)
C
C     COUNT ONLY THOSE INTERACTIONS WITH IST LESS THAN OR EQUAL TO JST
      IF((JST.EQ.0).OR.(IST.GT.JST)) GO TO 6000
C
C     CHECK TO SEE IF THIS IS A SPECIFIC INTERACTION IN THE TABLE
      IF(KSF(IST,JST).EQ.0) GO TO 600U
C
C     CHECK TO SEE IF THIS IS BIPHENYLENE-BIPHENYLENE INTERACTION.
C     IF IT IS, AT IS COUNTED ONLY ONCE AND THE UNIQUE ATOM MUST BE DESIGNATED.
      IF((KSF(IST,JST).EQ.31).AND.(NSTOP.EQ.0)) GO TO 6028
      IF((KSF(IST,JST).EQ.31).AND.(NSTOP.NE.0)) GO TO 6000
C
C     CHECK TO SEE IF THIS IS A RING INTERACTION (TYPE 23 OR 24)  COUNT
C     THE INTERACTION ONLY IF THE ATOMS ARE ON THE SAME RING.
      A = KSF(IST,JST)
      IF((A.EQ.23).OR.(A.EQ.24)) GO TO 850
      GO TO 851
850   DO 852 II=1,NR
      A = JR(II) + KR(II)
      B = I + J
      IF(((I.EW.JR(II)).OR.(I.EQ.KR(II))).AND.(A.EW.B)) GO TO 851
852   CONTINUE
      GO TO 6000
851   CONTINUE
C     CHECK TO AVOID COUNTING INTERACTION TWICE WHEN  IST = JST
      IF((IST.EQ.JST).AND.(MS.NE.0)) GO TO 2060
```

```
C    COUNT THE INTERACTION
2070 MS = MS + 1
C    ASSIGN INTERACTION NUMBER
     MSN(MS) = KSF(IST,JST)
C
C    ASSIGN THE ATOMS INVOLVED IN THE INTERACTION
     ISI(MS) = I
     JSI(MS) = J
     GO TO 6000
C
C    IF THE INTERACTION IS BIS-BIPHENYLENE-BIS-BIPHENYLENE, FIND THE
C    UNIQUE ATOM.
6020 DO 6030 K = 1,5
     IF((K.EQ.I).OR.(K.EQ.J).OR.(ISTER(K).EQ.17)) GO TO 6030
     MSN(MS) = KSF(IST,JST)
     ISI(MS) = K
     JSI(MS) = 0
     NSTOP = NSTOP + 1
6030 CONTINUE
     GO TO 6000
C
C    IF THE INTERACTION INVOLVES THE SAME TWO ATOMS AS A PREVIOUS INTERACTION
C    IT SHOULD NOT BE COUNTED AGAIN.
2060 DO 2050 KK=1,MS
     IF((ISI(KK).EQ.J).AND.(JSI(KK).EQ.I)) GO TO 6000
2050 CONTINUE
C
C    THE INTERACTION CAN BE COUNTED
     GO TO 2070
6000 CONTINUE
5000 CONTINUE
17   CONTINUE
C
C    PRINT OUT INITIAL DATA
     IF(N.NE.50) GO TO 20
     GO TO 25
2906 PRINT 906
906  FORMAT(/*    ISOMERS TO BE CALCULATED*)
     GO TO 21
25   PRINT 907
```

```
907 FORMAT(/*    ALL ISOMERS ARE TO BE CALCULATED*)
    GO TO 30
21  PRINT 705,(ID(I),I = 1,N)
705 FORMAT(1X,8A10)
30  IF(NR.EQ.0) GO TO 50
    PRINT 915,NR
915 FORMAT(/*    THERE ARE*,I5,*RINGS IN THE MOLECULE*/)
    PRINT 916
916 FORMAT(*    TYPE    ATOM(S) INVOLVED*)
    PRINT 917,(IRTYPE(I),JR(I),KR(I),I = 1,NR)
50  CONTINUE
    IF(M.EQ.0) GO TO 70
    PRINT 920,M
920 FORMAT(/*    THERE ARE*,I5,* PI EFFECTS CONSIDERED*)
    PRINT 916
    DO 700 I=1,M
700 PRINT 917,IPI(I),IJ(I)
70  CONTINUE
    IF(MS.EQ.0) GO TO 90
    PRINT 925,MS
925 FORMAT(/*    THERE ARE*,I5,*  STERIC EFFECTS CONSIDERED*)
    PRINT 916
    DO 861 I = 1,MS
    IF(MSN(I).EQ.31) GO TO 862
    PRINT 917,MSN(I),ISI(I),JSI(I)
    GO TO 861
862 PRINT 555,MSN(I),ISI(I)
861 CONTINUE
555 FORMAT(3(5X,I5,5X))
90  FORMAT(5X,I5,5X,I5,*  ATOM NOT BONDED TO BIS-BIPHENYLENE*)
    CONTINUE
    IF(NSE.NE.2) GO TO 2003
    PRINT 2004
2004 FORMAT(/*    GROUPS USED TO CALCULATE STERIC INTERACTIONS*)
2005 PRINT 2005
     FORMAT(*      ATOM NUMBER          GROUP*)
     DO 2006 I = 1,15
2006 PRINT 2007,I,ISTER(I)
2007 FORMAT(10X,I5,15X,I5)
2003 IF(MFU.GT.0) PRINT 674
674  FORMAT(/*  0.75 FACTOR APPLIED TO STERIC INTERACTIONS.  SEE FOOTN
    1OTE A, TABLE 5.9*)
C
C PRINT HEADINGS FOR ISOMER ENERGIES
```

```
      PRINT 739
      PRINT 740
  739 FORMAT(///* ISOMER  ISOMER ENERGY  ( KCAL/MOLE)
  740 FORMAT(*  ISOMER  SUBSTITUENT EFFECT  DELTA *)
     1EFFECT*,/,RING STRAIN  P-5 EFFECT  TOTAL ENERGY  STERIC
     1CTIONS*,/,115X,*WITH 0.75 FACTOR*)  STERIC ENERGY  STERIC INTERA
C
C ISOMER LOOP BEGINS.  K GOVERNS ISOMER LOOP.
C
      DO 300 K=1,N
      IF(ID(K).NE.4HTP12) GO TO 102
      INO=0
  102 IF(ID(K).NE.4HTP21) GO TO 104
      INO=5
  104 IF(ID(K).NE.4HTP13) GO TO 106
      INO=10
  106 IF(ID(K).NE.4HTP31) GO TO 108
      INO=15
  108 IF(ID(K).NE.4HTP14) GO TO 110
      INO=20
  110 IF(ID(K).NE.4HTP41) GO TO 112
      INO=25
  112 IF(ID(K).NE.4HTP15) GO TO 114
      INO=30
  114 IF(ID(K).NE.4HTP51) GO TO 116
      INO=35
  116 IF(ID(K).NE.4HTP23) GO TO 118
      INO=40
  118 IF(ID(K).NE.4HTP32) GO TO 120
      INO=45
  120 IF(ID(K).NE.4HTP24) GO TO 122
      INO=50
  122 IF(ID(K).NE.4HTP42) GO TO 124
      INO=55
  124 IF(ID(K).NE.4HTP25) GO TO 126
      INO=60
  126 IF(ID(K).NE.4HTP52) GO TO 128
      INO=65
  128 IF(ID(K).NE.4HTP34) GO TO 130
      INO=70
  130 IF(ID(K).NE.4HTP43) GO TO 132
      INO=75
  132 IF(ID(K).NE.4HTP35) GO TO 134
      INO=80
```

```
  134    IF(ID(K).NE.4HTP53) GO TO 136
         INO = 85
  136    IF(ID(K).NE.4HTP45) GO TO 138
         INO = 90
  138    IF(ID(K).NE.4HTP54) GO TO 140
         INO = 95
  140    IF(ID(K).NE.7HSP12345) GO TO 142
         INO = 100
  142    IF(ID(K).NE.7HSP12543) GO TO 144
         INO = 105
  144    IF(ID(K).NE.7HSP12435) GO TO 146
         INO = 110
  146    IF(ID(K).NE.7HSP12534) GO TO 148
         INO = 115
  148    IF(ID(K).NE.7HSP12354) GO TO 150
         INO = 120
  150    IF(ID(K).NE.7HSP12453) GO TO 152
         INO = 125
  152    IF(ID(K).NE.7HSP21345) GO TO 154
         INO = 130
  154    IF(ID(K).NE.7HSP21543) GO TO 156
         INO = 135
  156    IF(ID(K).NE.7HSP21435) GO TO 158
         INO = 140
  158    IF(ID(K).NE.7HSP21534) GO TO 160
         INO = 145
  160    IF(ID(K).NE.7HSP21354) GO TO 162
         INO = 150
  162    IF(ID(K).NE.7HSP21453) GO TO 164
         INO = 155
  164    IF(ID(K).NE.7HSP31425) GO TO 166
         INO = 160
  166    IF(ID(K).NE.7HSP31524) GO TO 168
         INO = 165
  168    IF(ID(K).NE.7HSP31245) GO TO 170
         INO = 170
  170    IF(ID(K).NE.7HSP31542) GO TO 172
         INO = 175
  172    IF(ID(K).NE.7HSP31254) GO TO 174
         INO = 180
  174    IF(ID(K).NE.7HSP31452) GO TO 176
         INO = 185
  176    IF(ID(K).NE.7HSP41325) GO TO 178
         INO = 190
```

```
178   IF(ID(K).NE.7HSP41523) GO TO 180
      INO=195
180   IF(ID(K).NE.7HSP41235) GO TO 182
      INO=200
182   IF(ID(K).NE.7HSP41532) GO TO 184
      INO=205
184   IF(IC(K).NE.7HSP41253) GO TO 186
      INO=210
186   IF(IC(K).NE.7HSP41352) GO TO 188
      INO=215
188   IF(ID(K).NE.7HSP51324) GO TO 190
      INO=220
190   IF(ID(K).NE.7HSP51423) GO TO 192
      INO=225
192   IF(ID(K).NE.7HSP51234) GO TO 194
      INO=230
194   IF(ID(K).NE.7HSP51432) GO TO 196
      INO=235
196   IF(ID(K).NE.7HSP51243) GO TO 198
      INO=240
198   IF(ID(K).NE.7HSP51342) GO TO 200
      INO=245
200   CONTINUE
      DO 212 I=1,5
212   NTYPE(I)=NN(I+INO)
      IF(INO.LE.95) DELTA = 0.0
      IF(INO.GT.95) DELTA = 7.0
C
C     CALCULATE ENERGY FROM ELEMENT EFFECT
      TOT = 0.0
      DO 211 I=1,5
      TOT=ELMAT(LATOM(I),NTYPE(I)) + TOT
211   CONTINUE
      EELEM(K) = TOT
C
C     CALCULATE ENERGY FROM PI EFFECTS
      IF(M.EQ.0) GO TO 812
      TOT=0.0
      DO 213 I=1,M
      EPI(K)=API(IPI(I),NTYPE(IJ(I))) + TOT
213   CONTINUE
      GO TO 214
812   EPI(K)=0.0
214   CONTINUE
```

```fortran
C  CALCULATE ENERGY FROM STERIC EFFECTS
      TOT = 0.0
      IF(MS.EQ.0) GO TO 216
      FAC = 1.0
      DO 217 I=1,MS
      NQ = NTYPE(ISI(I)) + NTYPE(JSI(I))
      P = 0
      IF(NQ.EQ.3) P = 1
      IF(NQ.EQ.9) P = 1
      IF(NQ.EQ.7) P = 1
      IF((NQ.EQ.8).AND.(NTYPE(ISI(I)).NE.NTYPE(JSI(I)))) P = 1
      IF((MSN(I).EQ.23).OR.(MSN(I).EQ.24).AND.(NQ.NE.3)) P = 0
C  CHECK TO SEE IF INTERACTION IS BIS-BIPHENYLENE IN 4 BASAL POSITIONS
C  OF SP
      IF(MSN(I).EQ.31) GO TO 2015
      IF(P.EQ.0) GO TO 215
C  SEE IF SPECIAL FACTOR IS TO BE USED FOR CERTAIN INTERACTIONS
      IF(MFU.EQ.0)GOTO 606
C  SEE IF THE INTERACTION IS AX-EQ OR AP-BAS
      IF(P.EQ.1).AND.(NQ.NE.9)) GO TO 618
C  FIND OUT IF THERE IS AN INTERACTION IDENTICAL TO THE I INTERACTION
  618 DO 607 J = 1,MS
      IF(MSN(J).NE.MSN(I)) GO TO 607
      NJ = NTYPE(ISI(J)) + NTYPE(JSI(J))
C  SEE IF THE IDENTICAL INTERACTION IS ALSO AX-EQ OR AP-BAS
      IF(NJ.EQ.NQ) GO TO 609
      NQJ = NJ + NQ
      IF(NQJ.EQ.15) GO TO 609
      GO TO 607
C  RULE OUT A TRANS BAS-BAS INTERACTION
  609 IF(NTYPE(ISI(J)).EQ.(NTYPE(JSI(J))) GO TO 607
C  STERIC INTERACTIONS  MAKE SURE IDENTICAL INTERACTIONS DON T INVOLVE TRANS
C  BAS ATOMS
      IF((NQ.EQ.7).AND.(NJ.EQ.7)) GO TO 607
      IF((NQ.EQ.8).AND.(NJ.EQ.8)) GO TO 607
C  FIND OUT IF THE AXIAL/APICAL ATOM IS INVOLVED IN OTHER INTERACTIONS
C  ONE OF THE TWO ATOMS INVOLVED IN INTERCCTION I IS AXIAL /APICAL
```

```
C     THE AXIAL/APICAL ATOM IS SET EQUAL TO A
      A = ISI(I)
      IF((NTYPE(A).EQ.2).OR.(NTYPE(A).EQ.3)) GO TO 608
      II = JSI(I)
608   DO 610 IL = 1,MS
      IF((IL.EQ.J).OR.(IL.EQ.I)) GO TO 610
      IF((ISI(IL).EQ.A).OR.(JSI(IL).EQ.A)) GO TO 607
610   CONTINUE
C     FIND OUT IF THE AXIAL/APICAL ATOM IS PART OF A RING
      IF(NR.EQ.0) GO TO 617
      DO 615 IL = 1,NR
      IF((JR(IL).EQ.A).OR.(KR(IL).EQ.A)) GO TO 607
615   CONTINUE
617   CONTINUE
C     FOR TP FIND OUT IF BOTH INTERACTIONS INVOLVE THE SAME AXIAL ATOM.
      NB = ISI(J)
      IF (NTYPE(A).NE.2) GO TO 671
      IF (NTYPE(NB).EQ.2) GO TO 860
      NB = JSI(J)
860   IF(NB.NE.A) GO TO 607
      FAC = 0.75
      PRINT 626, ID(K),ISI(I),JSI(I)
671   FORMAT(10X,A7,1X,*ATOM*,I2,1X,*AND*,I3)
626   CONTINUE
607   GO TO 606
C     CHECK TO SEE IF THE FOUR BIS-BIPHENYLENE ATOMS ARE IN BASAL POSITION
2015  IF(NTYPE(ISI(I)).NE.3) GO TO 215
606   TOT = (STFAC(MSN(I3)) + FAC) + TOT
      FAC = 1.0
215   TOT = TOT
217   CONTINUE
216   ASTERIC(K) = TOT
C     CALCULATE ENERGY FROM RING STRAIN EFFECTS
      TOT = 0.0
      IH = 0
      IF(NR.EQ.0) GO TO 225
      DO 226 I = 1,NR
      NQ = NTYPE(JR(I)) + NTYPE(KR(I))
      IF(NQ.EQ.3) P = 1
      IF(NQ.EQ.2) P = 2
```

```
      IF(NQ.EQ.4)IH = 1
      IF(NQ.EQ.10) IH = 2
      IF(NQ.EQ.8).AND.(NTYPE(JR(I)).NE.(NTYPE(KR(I)))) P = 4
      IF(NQ.EQ.7) P = 4
      IF(NQ.EQ.9) P = 3
      IF((NQ.EQ.8).AND.(NTYPE(JR(I)).EQ.(NTYPE(KR(I)))))IH=2
      IF(IH.GE.1)GO TO 228
      TOT=IRS((IRTYPE(I)),P) + TOT
226   CONTINUE
      ERSTRN(K) = TOT
      GO TO 230
225   CONTINUE
      ERSTRN(K) = 0.0
      GO TO 230
228   ERSTRN(K) = 1000.
230   CONTINUE
C
C     PRINT ENERGY FOR THE ISOMER
      IF(ERSTRN(K).EQ.1000.) GO TO 261
      ETOT(K) = EELEM(K)+ESTERIC(K) + ERSTRN(K) + EPI(K)+ DELTA
      PRINT 950,ID(K),EELEM(K),DELTA,ESTERIC(K),ERSTRN(K),EPI(K), ETOT(K)
950   FORMAT(5X,A10,6(5X,F10.5)/)
      GO TO 270
261   PRINT 958,ID(K),EELEM(K),DELTA,ESTERIC(K),EPI(K)
958   FORMAT(5X,A10,3(5X,F10.5),20X,F10.5)
      ETOT(K)=1000.
      PRINT 961
961   FORMAT(* TOTAL ENERGY NOT CALCULATED DUE TO RING LOCATION IN
     1 AX-AX OR TRANS BAS-BAS POSITION*/)
270   CONTINUE
300   CONTINUE
C
C     PRINT ISOMERS IN ORDER OF DECREASING ENERGY
      PRINT 982
982   FORMAT(/* ISOMER        ENERGY(KCAL/MOLE) */)
      DO 280 K = 1,N
      IF(ETOT(K).LT.1000.) GO TO 280
      PRINT 970,ID(K)
970   FORMAT(1X,A10,*              HIGH ENERGY*)
      ETOT(K) = -1.0
280   CONTINUE
296   EHIGH = ETOT(1)
      DO 290 K = 1,N
      IF(ETOT(K).GE.EHIGH) GO TO 310
      GO TO 290
```

```
310    EHIGH = ETOT(K)
       KK = K
290    CONTINUE
       WRITE(1,1000) (ETOT(K1),K1=1,N)
1000   FORMAT(1X,5(F10.5,1X))
       WRITE(1,1002) EHIGH,KK
1002   FORMAT(* EHIGH = *,F10.5,* KK= *,I5)
       IF(ETOT(KK).LT.0.0) GO TO 298
       PRINT983,ID(KK),ETOT(KK)
983    FORMAT(1X,A10,15X,F10.5)
       WRITE(1,1003) ID(KK),ETOT(KK)
1003   FORMAT(1X,020,1X,020)
       ETOT(KK) = -1.0
       GO TO 296
298    CONTINUE
C      RESTORE ID MATRIX VALUES IF N IS LESS THAN 50
       IF(N.EQ.50) GO TO 400
       DO 333 I=1,N
333    ID(I) = IDK(I)
C      CHECK TO SEE IF THERE IS ANOTHER MOLECULE TO BE CALCULATED
       GO TO 400
401    STOP
       END
```

Sample calculations follow for entries 8 and 17 listed in Table 5.11.

```
ENTRY 8     PF3Y2         (Y = N(CH3)2)                      N
F     0     0     2                          N
13    1     4     1                          F
13    13    5     1
ENTRY 17 = ENTRY18     R1 = OCH3   O    R2 = OPH   X=H   Y=PH   C
F     0     0     0                          N
8     1     0     5                          O
10    8     2     8
KX    0                                      C
```

TABLE OF SUBSTITUENT EFFECTS

SUBSTITUENT TYPE	ELECTRONEGATIVITY	TP(EQ)	TP(AX)	SP(AP)	SP(BAS)	SP(BAS)
1	-1.5	-10.5	14.5	-2.4	7.4	7.4
2	2.9	11.0	9.7	0.0	0.2	0.2
3	2.5	11.3	7.3	3.0	0.9	0.9
4	2.6	11.5	9.9	6.0	7.5	7.5
5	2.7	11.7	3.5	3.3	5.2	5.2
6	2.8	11.9	8.4	8.3	0.4	0.4
7	3.0	12.2	3.7	1.4	2.1	2.1
8	3.1	12.4	3.5	5.5	1.1	1.1
9	3.5	20.8	2.1	7.0	1.1	1.1
10	3.5	30.0	3.1	.	.	.
11	3.7	30.0	7.5	.	.	.
12	3.8	33.0	0.	0.	0	0

TABLE OF PI EFFECTS

LIGAND TYPE	TP(EQ)	TP(AX)	SP(AP)	SP(BAS)	SP(BAS)
1	0.0	6.0	6.0	6.0	6.0
2	0.0	2.0	2.0	2.0	2.0
3	0.0				

TABLE OF VALUES FOR STERIC INTERACTIONS APPLIED FOR TP(AX-EQ), SP(AP-BAS), SP(CIS BAS-BAS) INTERACTIONS

TYPE OF INTERACTION	FACTOR
1	6.5
2	6.0
3	6.0
4	6.0
5	5.5
6	5.0
7	4.5
8	4.0
9	3.5
10	3.5
11	3.0
12	3.0
13	3.0
14	2.5
15	2.0
16	2.0
17	2.0

RING STRAIN VALUES

RING TYPE	TP(AX-EQ)	TP(EQ-EQ)	SP(CIS BAS-BAS)	SP(AP-BAS)
1	8.0	24.0	2.1	16.0
2	2.2	15.8	1.1	.9
3	2.2	18.0	1.0	14.0
4	2.2	24.4	1.1	14.9
5	2.2	15.1	1.1	13.05
6	2.2	21.3	1.1	15.5
7	2.2	23.2	1.0	11.5
8	2.9	16.5	1.0	
9	2.9			
10				

RING TYPE	
18	0.0
19	0.0
20	0.0
21	1.1
22	1.1
23	1.3
24	2.0
25	1.0
26	8.8
27	6.9
28	6.5
29	5.4
30	0.7
31	6.6
32	3.3
33	
34	
35	
36	
37	0.00
38	2.00
39	3.0

```
ENTRY 8      PF3Y2      (Y = N(CH3)2)

ATOM NO.      SUBSTITUENT      SUBSTITUENT TYPE
1             F                12
2             F                12
3             F                12
4             Z                7
5             Z                7

ALL ISOMERS ARE TO BE CALCULATED

THERE ARE   2 PI EFFECTS CONSIDERED
TYPE       ATOM(S) INVOLVED
1          1          4
           1          5

THERE ARE   1 STERIC EFFECTS CONSIDERED
TYPE       ATOM(S) INVOLVED
1          1          4          5

GROUPS USED TO CALCULATE STERIC INTERACTIONS
ATOM NUMBER      GROUP
1                13
2                13
3                13
4                1
5                1

0.75 FACTOR APPLIED TO STERIC INTERACTIONS.   SEE FOOTNOTE a, TABLE 5.9
```

ISOMER ENERGY (KCAL/MOLE)

ISOMER	SUBSTITUENT EFFECT	DELTA	STERIC EFFECT	RING STRAIN	PI EFFECT	TOTAL ENERGY
TP12	7.40000	0.00000	0.00000	0.00000	0.00000	7.40000
TP21	7.40000	0.00000	0.00000	0.00000	0.00000	7.40000
TP13	7.40000	0.00000	0.00000	0.00000	0.00000	7.40000
TP31	7.40000	0.00000	0.00000	0.00000	0.00000	7.40000
TP14	11.00000	0.00000	6.50000	0.00000	6.00000	23.50000
TP41	11.00000	0.00000	6.50000	0.00000	6.00000	23.50000
TP15	11.00000	0.00000	6.50000	0.00000	6.00000	23.50000
TP51	11.00000	0.00000	6.50000	0.00000	6.00000	23.50000
TP23	7.40000	0.00000	0.00000	0.00000	0.00000	7.40000
TP32	7.40000	0.00000	0.00000	0.00000	0.00000	7.40000
TP24	11.00000	0.00000	6.50000	0.00000	6.00000	23.50000
TP42	11.00000	0.00000	6.50000	0.00000	6.00000	23.50000
TP25	11.00000	0.00000	6.50000	0.00000	6.00000	23.50000
TP52	11.00000	0.00000	6.50000	0.00000	6.00000	23.50000
TP34	11.00000	0.00000	6.50000	0.00000	6.00000	23.50000
TP43	11.00000	0.00000	6.50000	0.00000	6.00000	23.50000
TP35	11.00000	0.00000	6.50000	0.00000	6.00000	23.50000
TP53	11.00000	0.00000	6.50000	0.00000	6.00000	23.50000
TP45	14.60000	0.00000	0.00000	0.00000	12.00000	26.60000
TP54	14.60000	0.00000	0.00000	0.00000	12.00000	26.60000
SP12345	8.40000	7.00000	6.50000	0.00000	12.00000	33.90000
SP12543	8.40000	7.00000	6.50000	0.00000	12.00000	33.90000
SP12435	8.40000	7.00000	0.00000	0.00000	12.00000	27.40000
SP12534	8.40000	7.00000	0.00000	0.00000	12.00000	27.40000

SP12453	8.40000	7.00000	6.50000	0.00000	12.00000	33.90000
SP21345	8.40000	7.00000	6.50000	0.00000	12.00000	33.90000
SP21543	8.40000	7.00000	6.50000	0.00000	12.00000	33.90000
SP21435	8.40000	7.30000	0.00000	0.00000	12.00000	27.40000
SP21534	8.40000	7.00000	0.00000	0.00000	12.00000	27.40000
SP21354	8.40000	7.00000	6.50000	0.00000	12.00000	33.90000
SP21453	8.40000	7.00000	6.50000	0.00000	12.00000	33.90000
SP31425	8.40000	7.00000	0.00000	0.00000	12.00000	27.40000
SP31524	8.40000	7.00000	0.00000	0.00000	12.00000	27.40000
SP31245	8.40000	7.00000	6.50000	0.00000	12.00000	33.90000
SP31542	8.40000	7.00000	6.50000	0.00000	12.00000	33.90000
SP31254	8.40000	7.00000	6.50000	0.00000	12.00000	33.90000
SP31452	8.40000	7.00000	6.50000	0.00000	12.00000	33.90000
SP41325	5.00000	7.00000	6.50000	0.00000	12.00000	30.50000
SP41523	5.00000	7.00000	6.50000	0.00000	12.00000	30.50000
SP41235	5.00000	7.00000	6.50000	0.00000	12.00000	30.50000
SP41532	5.00000	7.00000	6.50000	0.00000	12.00000	30.50000
SP41253	5.00000	7.00000	6.50000	0.00000	12.00000	30.50000
SP41352	5.00000	7.00000	6.50000	0.00000	12.00000	30.50000
SP51324	5.00000	7.00000	6.50000	0.00000	12.00000	30.50000
SP51423	5.00000	7.00000	6.50000	0.00000	12.00000	30.50000
SP51234	5.00000	7.00000	6.50000	0.00000	12.00000	30.50000
SP51432	5.00000	7.00000	6.50000	0.00000	12.00000	30.50000
SP51243	5.00000	7.00000	6.50000	0.00000	12.00000	30.50000
SP51342	5.00000	7.00000	6.50000	0.00000	12.00000	30.50000

ISOMER	ENERGY(KCAL/MOLE)
SP31452	33.90000
SP31254	33.90000
SP31542	33.90000
SP31245	33.90000
SP21453	33.90000
SP21354	33.90000
SP21543	33.90000
SP21345	33.90000
SP12453	33.90000
SP12354	33.90000
SP12543	33.90000
SP12345	33.90000
SP51342	30.50000
SP51243	30.50000
SP51432	30.50000
SP51234	30.50000
SP51423	30.50000
SP51324	30.50000
SP41352	30.50000
SP41253	30.50000
SP41532	30.50000
SP41235	30.50000
SP41523	30.50000
SP41325	30.50000
SP31524	27.40000
SP31425	27.40000
SP21534	27.40000
SP21435	27.40000
SP12534	27.40000
SP12435	27.40000
TP54	26.60000
TP45	26.60000
TP53	23.50000
TP35	23.50000
TP43	23.50000
TP34	23.50000
TP52	23.50000
TP25	23.50000
TP42	23.50000
TP24	23.50000
TP51	23.50000
TP15	23.50000

```
TP41                    23.50000
TP14                    23.50000
TP32                     7.40000
TP23                     7.40000
TP31                     7.40000
TP13                     7.40000
TP21                     7.40000
TP12                     7.40000
```

ENTRY 17 = ENTRY18   R1 = OCH3   R2 = OPH   X=H   Y=PH

```
ATOM NO.   SUBSTITUENT   SUBSTITUENT TYPE
   1           C                9
   2           O                9
   3           O                9
   4           O                2
   5           C                5
```

ALL ISOMERS ARE TO BE CALCULATED

THERE ARE   1 RINGS IN THE MOLECULE

```
TYPE      ATOM(S) INVOLVED
  8             1            5
```

THERE ARE   6 STERIC EFFECTS CONSIDERED

```
TYPE      ATOM(S) INVOLVED
 21           3            2
 20           3            4
 21           4            2
 15           5            2
 16           5            3
 16           5            4
```

GROUPS USED TO CALCULATE STERIC INTERACTIONS

```
ATOM NUMBER       GROUP
    1               0
    2              10
    3               8
    4               8
    5               4
```

0.75 FACTOR APPLIED TO STERIC INTERACTIONS.   SEE FOOTNOTE a, TABLE 5.9

ISOMER ENERGY (KCAL/MOLE)

ISOMER	SUBSTITUENT EFFECT	DELTA	STERIC EFFECT	RING STRAIN	PI EFFECT	TOTAL ENERGY
TP12	8.20000	0.00000	4.00000	2.00000	0.00000	14.20000
TP21	8.20000	0.00000	4.00000	2.00000	0.00000	14.20000
TP13	8.20000	0.00000	4.00000	2.00000	0.00000	14.20000
TP31	8.20000	0.00000	4.00000	2.00000	0.00000	14.20000
TP14	8.20000	0.00000	4.00000	2.00000	0.00000	14.20000
TP41	8.20000	0.00000	4.00000	2.00000	0.00000	14.20000
TP15	16.70000	0.00000	6.00000		0.00000	

TOTAL ENERGY NOT CALCULATED DUE TO RING LOCATION IN AX-AX OR TRANS BAS-BAS POSITION

ISOMER	SUBSTITUENT EFFECT	DELTA	STERIC EFFECT	RING STRAIN	PI EFFECT	TOTAL ENERGY
TP51	16.70000	0.00000	6.00000		0.00000	

TOTAL ENERGY NOT CALCULATED DUE TO RING LOCATION IN AX-AX OR TRANS BAS-BAS POSITION

ISOMER	SUBSTITUENT EFFECT	DELTA	STERIC EFFECT	RING STRAIN	PI EFFECT	TOTAL ENERGY
TP23	8.20000	0.00000	6.00000	21.00000	0.00000	35.20000
TP32	8.20000	0.00000	6.00000	21.00000	0.00000	35.20000
TP24	8.20000	0.00000	6.00000	21.00000	0.00000	35.20000
TP42	8.20000	0.00000	6.00000	21.00000	0.00000	35.20000
TP25	16.70000	0.00000	6.00000	2.00000	0.00000	24.70000
TP52	16.70000	0.00000	6.00000	2.00000	0.00000	24.70000
TP34	8.20000	0.00000	6.00000	21.00000	0.00000	35.20000
TP43	8.20000	0.00000	6.00000	21.00000	0.00000	35.20000
TP35	16.70000	0.00000	6.00000	2.00000	0.00000	24.70000
TP53	16.70000	0.00000	6.00000	2.00000	0.00000	24.70000
TP45	16.70000	0.00000	6.00000	2.00000	0.00000	24.70000
TP54	16.70000	0.00000	6.00000	2.00000	0.00000	24.70000
SP12345	10.30000	7.00000	6.00000	13.00000	0.00000	36.30000
SP12543	10.30000	7.00000	6.00000	13.00000	0.00000	36.30000
SP12435	10.30000	7.00000	6.00000	13.00000	0.00000	36.30000

SP12534	10.30000	7.00000	6.00000	13.00000	0.00000	36.30000
SP12354	10.30000	7.00000	6.00000	13.00000	0.00000	36.30000
SP12453	10.30000	7.00000	6.00000	13.00000	0.00000	36.30000
SP21345	10.30000	7.00000	7.00000	1.00000	0.00000	25.30000
SP21543	10.30000	7.00000	7.00000	1.00000	0.00000	25.30000
SP21435	10.30000	7.00000	7.00000	1.00000	0.00000	25.30000
SP21534	10.30000	7.00000	7.00000	1.00000	0.00000	25.30000
SP21354	10.30000	7.00000	8.00000		0.00000	

TOTAL ENERGY NOT CALCULATED DUE TO RING LOCATION IN AX-AX OR TRANS BAS-BAS POSITION

SP21453	10.30000	7.00000	8.00000		0.00000	

TOTAL ENERGY NOT CALCULATED DUE TO RING LOCATION IN AX-AX OR TRANS BAS-BAS POSITION

SP31425	10.30000	7.00000	7.00000	1.00000	0.00000	25.30000
SP31524	10.30000	7.00000	7.00000	1.00000	0.00000	25.30000
SP31245	10.30000	7.00000	7.00000	1.00000	0.00000	25.30000
SP31542	10.30000	7.00000	7.00000	1.00000	0.00000	25.30000
SP31254	10.30000	7.00000	8.00000		0.00000	

TOTAL ENERGY NOT CALCULATED DUE TO RING LOCATION IN AX-AX OR TRANS BAS-BAS POSITION

SP31452	10.30000	7.00000	8.00000		0.00000	

TOTAL ENERGY NOT CALCULATED DUE TO RING LOCATION IN AX-AX OR TRANS BAS-BAS POSITION

SP41325	10.30000	7.00000	7.00000	1.00000	0.00000	25.30000
SP41523	10.30000	7.00000	7.00000	1.00000	0.00000	25.30000
SP41235	10.30000	7.00000	7.00000	1.00000	0.00000	25.30000
SP41532	10.30000	7.00000	7.00000	1.00000	0.00000	25.30000
SP41253	10.30000	7.00000	8.00000		0.00000	

TOTAL ENERGY NOT CALCULATED DUE TO RING LOCATION IN AX-AX OR TRANS BAS-BAS POSITION

SP41352	10.30000	7.00000	8.00000		0.00000	

TOTAL ENERGY NOT CALCULATED DUE TO RING LOCATION IN   AX-AX OR TRANS BAS-BAS POSITION

SP51324	1.60000	7.00000	8.00000	13.00000	0.00000	29.60000
SP51423	1.60000	7.00000	8.00000	13.00000	0.00000	29.60000
SP51234	1.60000	7.00000	8.00000	13.00000	0.00000	29.60000
SP51432	1.60000	7.00000	8.00000	13.00000	0.00000	29.60000
SP51243	1.60000	7.00000	8.00000	13.00000	0.00000	29.60000
SP51342	1.60000	7.00000	8.00000	13.00000	0.00000	29.60000

ISOMER	ENERGY(KCAL/MOLE)
TP15	HIGH ENERGY
TP51	HIGH ENERGY
SP21354	HIGH ENERGY
SP21453	HIGH ENERGY
SP31254	HIGH ENERGY
SP31452	HIGH ENERGY
SP41253	HIGH ENERGY
SP41352	HIGH ENERGY
SP12453	36.30000
SP12354	36.30000
SP12534	36.30000
SP12435	36.30000
SP12543	36.20000
SP12345	36.20000
TP43	35.20000
TP34	35.20000
TP42	35.20000
TP24	35.20000
TP32	35.20000
TP23	35.20000
SP51342	29.60000
SP51243	29.60000
SP51432	29.60000
SP51234	29.60000
SP51423	29.60000

article (70) are used instead, the agreement between the two sets of $\Delta G\neq$ values becomes $\pm$ 1.8 kcal/mol (see Footnote e to Table 2.1). Closer agreement is possible if a steric term is associated with substituents attached to the ring carbon atoms (Volume I, p 169ff).

In dealing with the perfluorophosphoranes that contain one or more chlorine ligands, some uncertainty exists regarding their structural assignment (Volume I, p 120). If we assume that the ligands in the ground-state trigonal bipyramids have the axial site preference, F > Cl(Br) > CF$_3$, as suggested by the NMR studies (Volume I, p 120), and that the element effect of the CF$_3$ group approximates that for the fluorine atom (Table 2.1), then calculated and experimental $\Delta G\neq$ values (90, 91, 92, 93) for intramolecular exchange of CF$_3$ groups can be compared. These results are listed in Table 1.13.

The pseudorotational pathway describes the isomer interchange **31 $\rightleftarrows$ 12** when Y = F or Cl. The ligand numbering follows the designation in Figure 1.13 with reference to the topological graph in Figure 1.11. The activation energy for equilibration of CF$_3$ groups by either of the routes: **31 $\rightleftarrows$ 54 $\rightleftarrows$ 12** or **31 $\rightleftarrows$ 52 $\rightleftarrows$ 14 $\rightleftarrows$ 35 $\rightleftarrows$ 12**, is the same for these perfluoro derivatives under the model approximation, except for Entry **2** for which the pathway **31 $\rightleftarrows$ 54 $\rightleftarrows$ 12** has a slightly lower energy. All TP and SP isomers are included in the calculation.

For (Y) ligands other than F and Cl, ligand exchange in the tris(trifluoro) compounds takes place between the ground-state isomers, **52 $\rightleftarrows$ 35**. For Entries **3–6**, the exchange route, **52 $\rightleftarrows$ 31 $\rightleftarrows$ 54 $\rightleftarrows$ 12 $\rightleftarrows$ 35**, has a slightly lower activation energy according to the model calculation. When Z = F instead of CF$_3$ (Entries **10** and **11**), again exchange occurs between

Table 1.13. Activation Energies for Intramolecular Ligand Exchange for Trifluoromethylphosphoranes

	Ground-State Isomer $PXYZ_3$			$\Delta G^{\neq}$, kcal/mol	
Entry	X	Y	Z	Calcd[a]	Obsd[b]
1	CH_3	F	CF_3	$31 \rightleftarrows 12$	11.0	9.6
2	CH_3	Cl	CF_3	$31 \rightleftarrows 12$	9.4	11.1
3	CH_3	OCH_3	CF_3	$52 \rightleftarrows 35$	12.6	12.9
4	CH_3	SCH_3	CF_3	$52 \rightleftarrows 35$	16.9	15.5
5	CH_3	$N(CH_3)_2$	CF_3	$52 \rightleftarrows 35$	19.0	16.5
6	CH_3	CH_3	CF_3	$52 \rightleftarrows 35$	18.0	>17.2
7	$N(CH_3)_2$	F	CF_3	$31 \rightleftarrows 12$	12.0	12.2
8	$N(CH_3)_2$	Cl	CF_3	$31 \rightleftarrows 12$	10.4	9.8
9	SCH_3	F	CF_3	$31 \rightleftarrows 12$	9.9	11.5[c]
10	CH_3	CF_3	F	$52 \rightleftarrows 35$	11.0	9.4[d]
11	SCH_3	CF_3	F	$52 \rightleftarrows 35$	9.9	12.8[e]

[a] For entries involving the SCH_3 group, the element effect obtained from Table 2.1 was based on an assinged electronegativity of 2.5. Further, 2 kcal/mol associated with a loss of π bonding was applied when the thio group was not located at an equatorial site of a trigonal bipyramid (cf. p 37).
[b] Activation energies at 298°C for Entries **1–5** are from Ref *90*. Those for Entries **6–8** are from Ref *91*.
[c] Ref *92*.
[d] Ref *93*.

the ground-state structures **52** $\rightleftarrows$ **35**, with equal energies calculated for the two pseudorotational routes connecting these isomers.

The comparison of experimental with calculated exchange barriers for this series of compounds shows that they correlate to within ±1.2 kcal/mol. This represents the average deviation to which the model reproduced $\Delta G^{\neq}$ values on the more extensive series of cyclic and acyclic derivatives listed in Table 1.11.

Literature Cited

1. Hoffmann, R., Howell, J. M., Muetterties, E. L., *J. Am. Chem. Soc.* (1972) **94**, 3047.
2. Holmes, R. R., *Acc. Chem. Res.* (1972) **5**, 296.
3. Strich, A., Veillard, A., *J. Am. Chem. Soc.* (1973) **95**, 5574.
4. Kimball, G. E., *J. Chem. Phys.* (1940) **8**, 188.
5. Cotton, F. A., *J. Chem. Phys.* (1961) **35**, 228.
6. Craig, D. P., Maccoll, A., Nyholm, R. S., Orgel, L. E., Sutton, L. E., *J. Chem. Soc.* (1954) 332.
7. Pauling, L., "The Nature of the Chemical Bond," 3rd ed., Cornell, Ithaca, 1960.
8. Duffey, G. H., *J. Chem. Phys.* (1949) **17**, 196.
9. Volkov, V. M., Levin, A. A., Dyatkina, M. E., *Dokl. Akad. Nauk SSSR* (1963) **152**, 359.
10. Volkov, V. M., Levin, A. A., Dyatkina, M. E., *Soviet Phys. Dokl.* (1963) **152**, 804.
11. Brockway, L. O., Beach, J. Y., *J. Am. Chem. Soc.* (1938) **60**, 1836.

12. Rundle, R. E., *Rec. Chem. Progr.* (1962) **23**, 195.
13. Rundle, R. E., *J. Am. Chem. Soc.* (1963) **85**, 112.
14. Rundle, R. E., *Acta Crystallogr.* (1961) **14**, 585.
15. Rundle, R. E., *Surv. Progr. Chem.* (1963) **1**, 81.
16. Bartell, L. S., Hansen, K. W., *Inorg. Chem.* (1965) **4**, 1777.
17. Mellish, C. E., Linnett, J. W., *Trans. Faraday Soc.* (1954) **50**, 657.
18. Bent, H. A., *Can. J. Chem.* (1960) **38**, 1235.
19. Gillespie, R. J., Nyholm, R. S., *Quart. Rev. (London)* (1957) **11**, 339.
20. Gillespie, R. J., *J. Chem. Educ.* (1963) **40**, 295.
21. Gillespie, R. J., *J. Am. Chem. Soc.* (1963) **85**, 4671.
22. Gillespie, R. J., *Can. J. Chem.* (1960) **38**, 818.
23. Gillespie, R. J., *J. Chem. Soc.* (1963) 4672.
24. Gillespie, R. J., *Can. J. Chem.* (1961) **39**, 318.
25. Zemann, J., *Z. Anorg. Allegem. Chem.* (1963) **324**, 241.
26. Kepert, D. L., *Inorg. Chem.* (1973) **12**, 1938, 1942.
27. Gillespie, R. J., *Inorg. Chem.* (1966) **5**, 1634.
28. Bartell, L. S., *Inorg. Chem.* (1966) **5**, 1635.
29. van der Voorn, P. C., Drago, R. S., *J. Am. Chem. Soc.* (1966) **88**, 3255.
30. Berry, R. S., Tamres, M., Ballhausen, C. J., Johansen, H., *Acta Chem. Scand.* (1968) **22**, 231.
31. Holmes, R. R., Carter, R. P., Jr., Peterson, G. E., *Inorg. Chem.* (1964) **3**, 1748.
32. Carter, R. P., Jr., Holmes, R. R., *Inorg. Chem.* (1965) **4**, 738.
33. Holmes, R. R., *J. Chem. Phys.* (1967) **46**, 3718.
34. Maryott, A. A., Kryder, S. J., Holmes, R. R., *J. Chem. Phys.* (1965) **43**, 2556.
35. Mulliken, R. S., *J. Chem. Phys.* (1955) **23**, 1833, 1841, 2338, 2343.
36. Bartell, L. S., Plato, V., *J. Am. Chem. Soc.* (1973) **95**, 3097.
37. Clementi, E., Raimondi, D. L., *J. Chem. Phys.* (1963) **38**, 2686.
38. Bartell, L. S., Su, L. S., Yow, H., *Inorg. Chem.* (1970) **9**, 1903.
39. Berry, R. S., *J. Chem. Phys.* (1960) **32**, 933.
40. Holmes, R. R., Couch, L. S., Hora, C. J., Jr., *Chem. Commun.* (1974) 175.
41. Bernstein, L. S., Abramowitz, S., Levin, I. W., *J. Chem. Phys.* (1976) **64**, 3228.
42. Bernstein, L. S., Kim, J. J., Pitzer, K. S., Abramowitz, S., Levin, I. W., *J. Chem. Phys.* (1975) **62**, 3671.
43. Demuynck, J , Strich, A., Veillard, A., *Nouv. J. Chim.* (1977) **1**, 217.
44. Altmann, J. A., Yates, K., Csizmadia, I. G., *J. Am. Chem. Soc.* (1976) **98**, 1450.
45. Rauk, A., Allen, L. C., Mislow, K., *J. Am. Chem. Soc.* (1972) **94**, 3035.
46. Gillespie, P., Hoffmann, P., Klusacek, H., Marquarding, D., Pfohl, S., Ramirez, F., Tsolis, E. A., Ugi, I. *Angew. Chem. Int. Ed. Engl.* (1971) **10**, 687.
47. Ugi, I., Marquarding, D., Klusacek, H., Gillespie, P., Ramirez, F., *Acc. Chem. Res.* (1971) **4**, 288.
48. Florey, J. B., Cusachs, L. C., *J. Am. Chem. Soc.* (1972) **94**, 3040.
49. Santry, D. P., Segal, G. A., *J. Chem. Phys.* (1967) **47**, 158.
50. Brown, R. D., Peel, J. B., *Aust. J. Chem.* (1968) **21**, 2589, 2605, 2617.
51. Strich, A., *Inorg. Chem.* (1978) **17**, 942.
52. Musher, J., *J. Am. Chem. Soc.* (1972) **94**, 5662.
53. Dalton, B. J., *J. Chem. Phys.* (1971) **54**, 4745.
54. Whitesides, G. M., Mitchell, H. L., *J. Am. Chem. Soc.* (1969) **91**, 5384.
55. Muetterties, E. L., Meakin, P., Hoffmann, R., *J. Am. Chem. Soc.* (1972) **94**, 5674.
56. Muetterties, E. L., Mahler, W., Packer, K. J., Schmutzler, R., *Inorg. Chem.* (1964) **3**, 1298.
57. Mislow, K., *Acc. Chem. Res.* (1970) **3**, 321.
58. DeBruin, K. E., Naumann, K., Zon, G., Mislow, K., *J. Am. Chem. Soc.* (1969) **91**, 7031.
59. Gorenstein, D., Westheimer, F. H., *J. Am. Chem. Soc.* (1970) **92**, 634.

60. Cram, D. J., Day, J., Rayner, D. R., von Schriltz, D. M., Duchamp, D. J., Garwood, D. C., *J. Am. Chem. Soc.* (1970) **92**, 7369.
61. Luckenbach, R., "Dynamic Stereochemistry of Pentacoordinated Phosphorus and Related Elements," p. 214, Georg Thieme, Stuttgart, 1973.
62. Gielen, M., Dehouck, C., Mokhtar-Jamai, H., Topart, J., *Rev. Silicon Germanium, Tin, Lead Compds.* (1972) **1**, 9.
63. Lauterbur, P. C., Ramirez, F., *J. Am. Chem. Soc.* (1968) **90**, 6722.
64. Dunitz, J. D., Prelog, V., *Angew. Chem.* (1968) **80**, 700; *Angew. Chem., Int. Ed. Engl.* (1968) **7**, 725.
65. Muetterties, E. L., *J. Am. Chem. Soc.* (1969) **91**, 1636.
66. *Ibid.*, p. 4115.
67. King, R. B., *J. Am. Chem. Soc.* (1969) **91**, 7211.
68. Balaban, A. T., *Rev. Roum. Chim.* (1970) **15**, 432, 1257.
69. Gielen, M., "Chemical Applications of the Graph Theory," A. T. Balaban, Ed., Academic Press, New York, 1972.
70. Holmes, R. R., *J. Am. Chem. Soc.* (1978) **100**, 433.
71. Russegger, P., Brickmann, J., *Chem. Phys. Lett.* (1975) **30**, 276.
72. Eisenhut, M., Mitchell, H. L., Traficante, D. D., Kaufman, R. J., Deutch, J. M., Whitesides, G. M., *J. Am. Chem. Soc.* (1974) **96**, 5385.
73. Cowley, A. H., Braun, R. W., Gilje, J. W., *J. Am. Chem. Soc.* (1975) **97**, 434.
74. Mahler, W., Muetterties, E. L., *Inorg. Chem.* (1965) **4**, 1520.
75. Gilje, J. W., Braun, R. W., Cowley, A. H., *J. Chem. Soc., Chem. Commun.* (1974) 15.
76. Moreland, C. G., Doak, G. O., Littlefield, L. B., Walker, N. S., Gilje, J. W., Braun, R. W., Cowley, A. H., *J. Am. Chem. Soc.* (1976) **98**, 2161.
77. Gorenstein, D., *J. Am. Chem. Soc.* (1970) **92**, 664.
78. Bone, S. A., Trippett, S., White, M. W., Whittle, P. J., *Tetrahedron Lett.* (1974) 1795.
79. Bone, S., Trippett, S., Whittle, P. J., *J. Chem. Soc., Perkin Trans. 1* (1974) 2125.
80. White, D. W., De'Ath, N. J., Denney, D. Z., Denney, D. B., *Phosphorus* (1971) **1**, 91.
81. Denney, D. Z., White, D. W., Denney, D. B., *J. Am. Chem. Soc.* (1971) **93**, 2066.
82. Duff, R. E., Oram, R. K., Trippett, S., *Chem. Commun.* (1971) 1011.
83. Oram, R. K., Trippett, S., *J. Chem. Soc., Perkin Trans. 1* (1973) 1300.
84. Harris, R. K., Woplin, J. R., Dunmur, R. E., Murray, M., Schmutzler, R., *Ber. Bunsenges. Phys. Chem.* (1972) **76**, 44.
85. Hellwinkel, D., *Chemia (Aarau)* (1968) **22**, 488.
86. Whitesides, G. M., Eisenhut, M., Bunting, W. M., *J. Am. Chem. Soc.* (1974) **96**, 5398.
87. Hellwinkel, D., Wilfinger, H. J., *Tetrahedron Lett.* (1969) 3423.
88. Szobota, J. S., Holmes, R. R., *Inorg. Chem.* (1977) **16**, 2299.
89. Howard, J. A., Russell, D. R., Trippett, S., *J. Chem. Soc., Chem. Commun.* (1973) 856.
90. Cavell, R. G., Gibson, J. A., The, K. I., *Inorg. Chem.* (1978) **17**, 2880.
91. Cavell, R. G., personal communication.
92. Cavell, R. G., The, K. I., Gibson, J. A., Yap, N. T., *Inorg. Chem.* (1979) **18**, 3400.
93. Yap, N. T., Cavell, R. G., *Inorg. Chem.* (1979) **18**, 1301.

Chapter

2

Reaction Mechanisms

IN DISCUSSING THE reactions of phosphorus compounds, we focus on those transformations that seem to give evidence for the presence of pentacoordinated intermediates or transition states. The presentation is illustrative rather than exhaustive, and the reactions chosen are those for which the accumulation of evidence is most convincing in suggesting the necessity for invoking five-coordinate species in accounting for product formation and reaction kinetics. These include two widely studied classes of reactions: the hydrolysis of phosphate esters and the decomposition of phosphonium salts. Each class is represented by studies that use cyclic and acyclic derivatives and follows mechanistic sequences that are based on structural principles derived for pentacoordinated phosphorus compounds. When a particular mechanistic sequence seems to be at variance with the structural principles or seems to be less satisfactory in providing a self-consistent treatment, alternative mechanistic rationalizations are given. Considering the degree of uncertainty in defining most of the mechanisms outlined here, a proper choice in difficult cases most likely will become apparent as future work unfolds.

In a subsequent section, more complex reactions of phosphorus are considered. These center on biologically important processes, including ribonuclease catalysis, DNA replication, and RNA transcription. However, before we discuss any of the aforementioned processes, it is advantageous to summarize the main criteria to be applied in constructing an appropriate mechanistic scheme.

Typically, the trigonal bipyramid is the form postulated as the initial reaction intermediate in nucleophilic attack at tetracoordinated phosphorus. In the process, it is presumed that the entering and leaving groups enter and leave from axial positions. The latter is in accord with the principle of microscopic reversibility as phrased by Westheimer (1). Further, since the axial bonds are longer and apparently weaker than equatorial bonds (Volume I, p 246), the process has reasonable merit.

Depending on the lifetime of the intermediate and the associated degree of fluxional character, intramolecular ligand exchange can be rapid enough to be a primary factor in determining the course of the reaction. Passage through a square pyramidal transition state is implicit in this pseudorotational process if the mechanistic interpretation is based specifically on the Berry

0065–7719/80/0175–0087$31.60/1
© 1980 American Chemical Society

exchange process (2). Since the evidence for this exchange process appears compelling (*see* Volume I, p 110), we adopt it here as a convenient model.

Thus, to outline phosphorus reaction mechanisms properly, it is necessary to consider both trigonal bipyramidal and square pyramidal conformations and their relative activation energies.

Formation of Activated States

Preference Rules. In the preceding volume dealing with structural phenomena, certain principles emerged that govern conformational preferences regarding the positioning of ligands in a trigonal bipyramidal framework. A summary of the more apparent ones should aid in outlining their use in the important area of reaction mechanisms. They are listed as follows in order of probable importance.

1. Four- or five-membered cyclic systems preferentially span axial–equatorial positions.

2. The most electronegative ligands preferentially occupy axial sites.

3. Pi-bonding donor ligands, in general, are positioned at equatorial sites.

4. Steric effects are minimized by locating bulky groups in equatorial positions.

Based on theory and more limited experimental observation (Volume I, pp 10 and 11), similar criteria can be developed for square pyramidal structures. These are listed below and apply to a square pyramid having an apical–basal angle of ca. 105°. Less certainty is associated with the indicated order of importance.

1. Four- or five-membered cyclic systems preferentially span cis basal positions.

2. The most electronegative ligands preferentially occupy basal site.

3. Pi-bonding donor ligands, in general, are positioned at the apical site.

4. Steric effects are minimized by locating a bulky group in the apical position.

In arriving at the latter set of preference rules, use is made of the correspondence that exists between the positions in each of the two pentacoordinate geometries (*see* Volume I, p 116). Restated here, axial bonds of a trigonal bipyramid correspond more nearly in properties to basal bonds of a square pyramid while equatorial bonds of a trigonal bipyramid are more related in character to the apical bond of a square pyramid, assuming that identical ligands are involved. This duality in bond properties allows us to estimate the relative importance of pseudorotational transition states if we also can account satisfactorily for the principal factors (3) that determine the respective energies of the isomeric conformations that are undergoing possible interchange.

Relative Isomer Energies. In this regard, it is frequently useful to obtain a numerical estimate of the relative energies of these structures to

Table 2.1. Relative Electronegativity and Substituent Effects (kcal/mol) in Pentacoordinate Phosphorus Species[a]

Substituent	χ	TP[b,c] eq	TP[b,c] ax	SP[b,c] ap	SP[b,c] bas
$\cdot F$; $\cdot CF_3$	4.0	3.0	0.0	6.0	0.0
$\overset{\displaystyle H^+}{\overset{\displaystyle \vert}{\cdot O}}-H$, $\overset{\displaystyle H^+}{\overset{\displaystyle \vert}{\cdot O}}-C\,(\text{ring})$, $\cdot O-SiCl_3$	3.8	3.0	0.5	5.7	0.1
$\cdot O-H$, $\cdot O-C\,(\text{ring})$, $\overset{\displaystyle H^+}{\overset{\displaystyle \vert}{\cdot O}}-CH_3$, $\overset{\displaystyle H^+}{\overset{\displaystyle \vert}{\cdot O}}-C_2H_5$	3.7	3.0	0.7	5.5	0.2
$\cdot O$[d]	3.5	2.8	1.3	5.1	0.4
$\cdot OCH_3$[e]	3.1	2.4	2.4	4.1	1.0
$\overset{\displaystyle \vert}{\cdot N}-$, $\cdot Cl$	3.0	2.2	2.8	3.8	1.2
$\cdot SiCl_3$, $\cdot CH_2Ph$	2.8	1.9	3.5	3.3	1.5
$\cdot Ph$	2.7	1.7	3.9	3.0	1.7
$\cdot H$, $\cdot O^-$, $\cdot SH$, $\cdot SR$	2.5	1.3	4.7	2.3	2.2
$\cdot CH_3$, $\cdot CH_2CH_3$ $\cdot C(CH_3)_3$, $\cdot C$[f] Odd election	2.0	0.0	7.0	0.0	4.0
Lone election pair	1.5	−1.6	9.9	−2.9	7.4

[a] For details of construction of a table of this type consult p 32. TP = trigonal bipyramid; SP = square pyramid. All of the element effects included in this table as well as the ring strain, steric, and π terms referenced in the following footnotes have been incorporated in the FORTRAN program ISOMER 2 listed starting on p 51.

[b] Differences in the values of the element effect for the same substituent located in the two types of positions of either the TP (ax–eq) or SP (ap–bas) provide relative apicophilicities for the respective conformation. For either the TP or SP, the more negative the difference, the greater the preference for the axial site. In accordance with the preference rules, the more electronegative ligands prefer the axial position of the TP while the less electronegative ligands prefer the apical position of the SP.

[c] As mentioned (p 35), a value of 7.0 kcal/mol must be added to the sum of these substituent values for a SP to account for the inherent instability of the SP relative to the TP. Also, both steric and ring strain terms must be included. These are listed in Tables 1.8, 1.9, and 1.10 on pp 36–40. Additional entries appearing subsequent to these tables are summarized on p 37.

[d] Use this entry for any other oxygen atom type not specifically mentioned.

[e] When discussing $\Delta G^{\ddagger}$ values for intramolecular exchange (p 41), the OCH$_3$ group was assigned the element effect associated with the oxygen atom (43). In discussing reaction intermediates, the substituent effect for the OCH$_3$ group associated with an electronegativity of 3.1 appears more applicable.

[f] Use this entry for any other carbon atom type not specifically listed.

decide the feasibility of a particular reaction pathway. The model presented in Chapter 1 (p 32) provides this kind of basis. For those who may be interested in this approach, we have compiled a list of some of the groups encountered in mechanistic sequences treated in this chapter. Table 2.1 shows group electronegativities, obtained from estimates available in the literature (4, 5); they are listed in descending order along with corresponding values of substituent effects for each of the two types of positions found in the trigonal bipyramid and square pyramid. The latter values follow from the relationship between electronegativity and element effect obtained in the model development. As described in Chapter 1 (p 32), these values, which are a measure of relative apicophilicity (axiophilicity is a synonymous expression as used by most authors (see Ref 99)), when properly summed with steric and ring strain effects, provide a quantitative expression of the preference rules. The appropriate scales to use for the latter effects are found on pp 36–40. Since these scales were originally derived from activation energies governing fluxional behavior for ground-state pentacoordinate derivatives, they do not adequately take into account the leaving ability and anion stabilization of the departing axial group or reactions that occur under unusual steric influences (see pp 118–122 for systems under apparent kinetic rather than thermo-dynamic control).

It must be remembered that any scale of this type is simplistic and is subject to continuous scrutiny as new data on reaction mechanisms are ob-tained. However, application of these substituent scales in their present form does provide a unifying treatment and allows a common basis for discussing exceptions to generally observed trends.

Accordingly, for some of the reaction mechanisms that form the topics of this chapter, numerical estimates of relative energies (kcal/mol), obtained from the use of these scales, are inserted below structural formulas of postu-lated pentacoordinate intermediates. The particular forms of the designated intermediates agree with the principles cited above and have substituent charges that give the lowest numerical estimate of isomer energy. When protonated intermediates are involved, for example, in a pseudorotational process, the substituents that are protonated may differ in each of the isomeric states—i.e., on going from the initial trigonal bipyramid through a transitory square pyramidal state to the pseudorotated trigonal bipyramid.

Reaction Intermediates and Transition States. In preceding chap-ters the terms intermediate and transition states are used somewhat inter-changeably in discussing the relative merits of barrier conformations postu-lated in intramolecular ligand exchange mechanisms. In this chapter, as already applied, the term intermediate is reserved for trigonal bipyramids that are postulated to appear in a reaction mechanism. The geometries of these states are considered to approximate closely those observed for stable penta-coordinate phosphorus compounds that are related in substituent composition. However, for the most part, the intermediate trigonal bipyramids have gone undetected in the many reactions in which they have been regarded as being of primary importance in explaining reaction rates and product distributions. The implication is that the lifetime of such states is short enough or the concentrations are low enough that observation is rendered impractical by ordinary means. In this context, high-energy trigonal bipyramids, envisioned to form in concerted S_N2 displacements, will be termed intermediates.

In contrast, the term transition state is reserved for postulated conformations that lead to these unstable trigonal bipyramidal intermediates whether they are from reactants or products or are formed as a result of intervening transformations, such as pseudorotations that are frequently invoked.

Typical conformations representing transition states are distortions of tetracoordinated phosphorus occuring on the way to initial intermediate states as a result of interaction with an attacking group and square pyramids postulated in pseudorotational transformations that are assumed to proceed via the Berry mechanism (2). In terms of a potential energy surface governing a reaction scheme, transition states are envisioned as appearing at the crests while the unstable trigonal bipyramidal intermediates are regarded as appearing at troughs. In general, the latter most likely are relatively shallow and of substantially higher energies than reactant or product states to agree with the lack of direct experimental verification for their existence. Both trigonal bipyramidal intermediates and transition states are shown in brackets to differentiate their nature from intermediates that can be isolated. Although, the terms trigonal bipyramid and square pyramid are used for convenience, related phosphorane compositions (Volume I, Chapter 2) bear some measure of distortion from idealized geometries. Further, the terms square pyramid, tetragonal pyramid, and rectangular pyramid are used interchangeably in this chapter.

For those reaction mechanisms which have pseudorotational processes postulated as an important component, the question arises as to whether a square pyramid or some distortion of it uniformly represents a barrier state under the assumed Berry exchange process. The fact that some spirocyclic derivatives have nearly square pyramidal structures (Volume I, p 11) suggests that ligand exchange via trigonal bipyramidal transition states is possible. The characterization of a series of structures, whose deformations closely follow the Berry exchange coordinate, extending from near trigonal bipyramidal to near square or rectangular pyramidal (6, 7), illustrates that the energy balance between the two isomeric states is delicate (Volume I, p 42). However, a pentacoordinate phosphorus compound has yet to be found near an idealized square pyramidal configuration when the substituents are simple or monocyclic (8) (Volume I, pp 12 and 22). Nonetheless, it would be premature to disregard this possibility.

Acid Hydrolysis of Cyclic Esters

Methylethylene Phosphate. A particularly striking application to mechanistic behavior—one that had been discussed somewhat before the principles outlined in the preceding section were adequately formulated—concerns the hydrolysis of strained cyclic phosphate esters. Five-membered cyclic phosphate esters (9, 10, 11, 12) hydrolyze much more rapidly (10^5–10^8 times) than their open chain analogs in either acid or base. Six-membered

Table 2.2. Relative Rates of Hydrolysis of Cyclic (k_c) and Acyclic (k_a) Phosphates, Phosphonates, and Phosphinates

Entry	Compound	Conditions	$k_c/k_a{}^a$	Ring Retention, %	Ref
1	(five-membered cyclic PO$_2$) Ba, subscript 2	base	10^7	0	10
2	methylethylene phosphate (cyclic O–P(=O)–OCH$_3$)	acid	10^6	30–50	16, 24
		base	10^6	0	1, 24
3	(cyclic PO$_2^-$ Li$^+$)	acid	5×10^4	0	12
		base	6×10^5	0	12
4	(cyclic P(=O)–OCH$_3$, carbon ring)	acid	10^6	0	16
		base	10^6	0	12
5	(cyclopentane P(=O)–OC$_2$H$_5$)	acid	1	100	17, 19
		base	2–4	—	17, 19
6	(cyclobutene P(=O)–OC$_2$H$_5$)	acid	1	—	19
		base	0.14	—	19
7	(cyclopropane P(=O)–OC$_2$H$_5$)	acid	3	—	19
		base	~60	—	19

^{a} The rate constant for potassium dimethylphosphate which does not differ much from that of trimethylphosphate (*23*) is 3×10^{-11} Lmol^{-1} sec^{-1} at 25°C and unit ionic strength.

(*13*) and seven-membered (*14, 15*) cyclic phosphates hydrolyze at rates more comparable with those of acyclic phosphates. Accompanying the rapid rate in the five-membered cyclic esters, the ring is opened in some cases (*16, 17*), while in others, an appreciable percentage of the product arises from hydrolysis of the ester group without ring opening (*18, 19, 20, 21, 22*). For methylethylene phosphate, ring retention in the range 30–50% is obtained in acid solution (*16, 18*) (Table 2.2).

Westheimer (*1, 16, 23, 24*) and co-workers have interpreted the wide difference in rates most reasonably by assuming that the hydrolysis involves a trigonal bipyramidal intermediate that lowers the strain in the cyclic ester. The mechanism proposed for methylethylene phosphate involves addition of a proton and a water molecule, leading to Intermediates **I** and **Ia**. In mildly acid solution, the equilibrium shifts toward **I**. Formation of **I** reduces the ring strain that exists in the initial phosphate ester. The preferred O–P–O angle in the five-membered ring is about 90° (*see* Volume I, p 22). However, in methylethylene phosphate, x-ray analysis (25) has shown this angle to be 99°. The latter is typical of tri- and tetracoordinated phosphorus compounds. The corresponding angle in $(CH_3)_3P$ (26), for example, is 99°. Therefore, it seems that methylethylene phosphate has approximately 10° of strain. Using

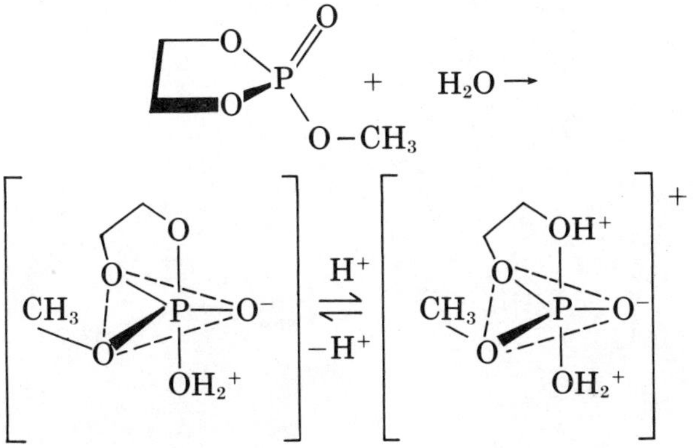

angle-bending force constants, Usher (27) et al. calculated a relief of strain of 3–6 kcal/mol associated with the formation of **I**. Additional stabilization of the cyclic intermediate most likely is derived from entropic effects (28).

Pseudorotation of **I** brings the methanol molecule into a leaving position. This takes place about the P–O equatorial linkage as a pivotal group passing through the square pyramidal transition state **II** to result in the trigonal bipyramidal intermediate **III**. Hydrolysis with ring opening may proceed directly from **Ia** without pseudorotation.

The pH rate profile (24) for the hydrolysis of methylethylene phosphate shows that the amount of ring retention decreases in strongly acidic solutions, i.e., pH <2. This is consistent with an expected increased concentration of the protonated form **Ia** if pseudorotation (as Westheimer and co-workers reason (24)) occurs predominantly by way of the unprotonated form, **I**. Thus, the acid-catalyzed rate of formation of **Ia**, and hence endocyclic cleav-

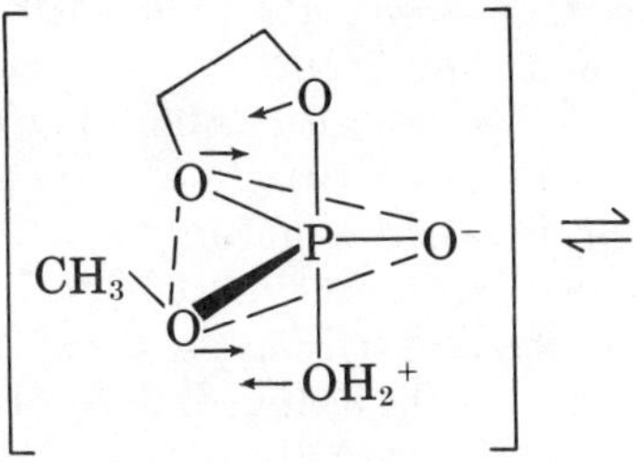

9.9

I

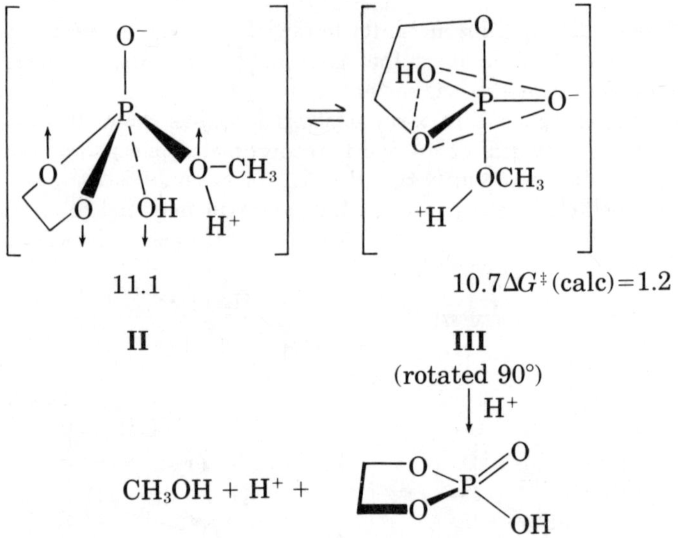

$$11.1 \qquad\qquad 10.7 \Delta G^{\ddagger}(\text{calc}) = 1.2$$

II   **III**

(rotated 90°)

$$\downarrow H^+$$

$$CH_3OH + H^+ +$$

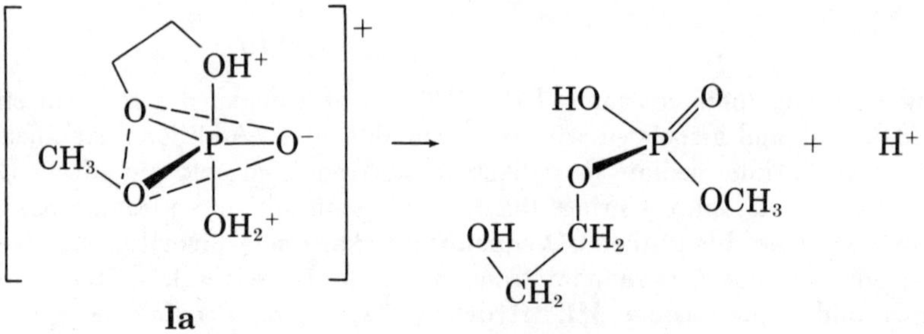

Ia

age, is enhanced in strong acid, while the pseudorotational process, **I** ⇄ **III**, becomes rate limiting. (We have introduced the charge separation in **I** and **Ia** to correspond to our stated condition (p 90) of distributing substituent charges in each isomer so that the lowest energy summation results from application of the substituent effects listed in Table 2.1. Under this scheme, we calculate a low pseudorotational barrier for **Ia** of 1.3 kcal/mol, comparable with that for **I**. However, the protonated form **Ia** contains a better endocyclic leaving group. More conventionally, the endocyclic cleavage of **I** may be viewed as a water-catalyzed reaction that is slow (*24*). Further detail is discussed by Luckenbach (*29*).

Methyl Ester of Propylphostonic Acid. Evidence that the pseudo-rotational process is necessary to the latter hydrolysis mechanism is supplied by the data (*16*) from the acid hydrolysis of the methyl ester of propylphos-tonic acid, **IV**. In contrast to the acid hydrolysis of methylethylene phosphate, the rapid hydrolysis of **IV** proceeds with little observable ring retention (Table 2.2). The intermediate trigonal bipyramid **V**, postulated to form initially in this case, places the more electronegative oxygen atom of the five-membered ring in an axial position, in agreement with the second rule cited for trigonal bipyramids (p 88).

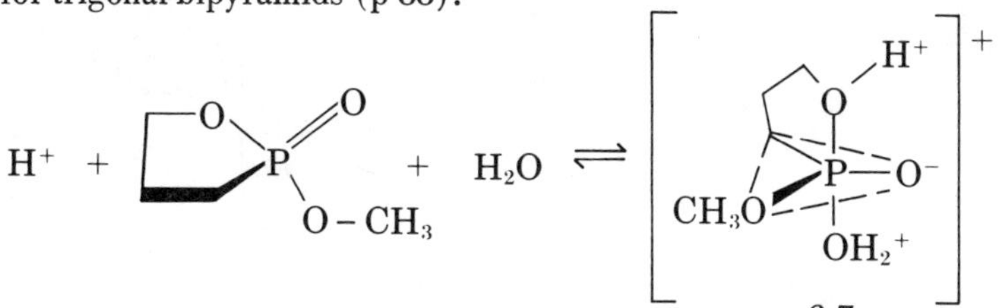

$$H^+ + \text{IV} + H_2O \rightleftharpoons \text{V} \quad 6.7$$

Dennis and Westheimer (*16*) argued that ring cleavage is expected since there is no low-energy pathway to place the methoxy group in an axial position from which it can leave. Pseudorotation about the equatorial methylene group introduces strain in the five-membered ring by bringing it in an equatorial position (**VI**) with a preferred phosphorus angle of 120° (contrary to rule 1). Likewise, pseudorotation about the equatorial P–O bond places the methylene group in an unfavorable axial position in violation of the electronegativity rule (**VII**).

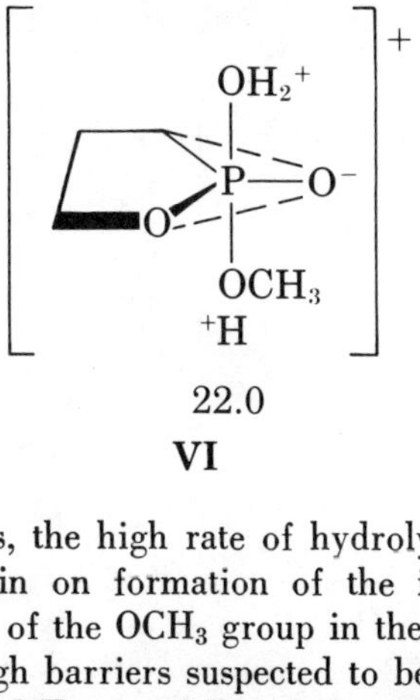

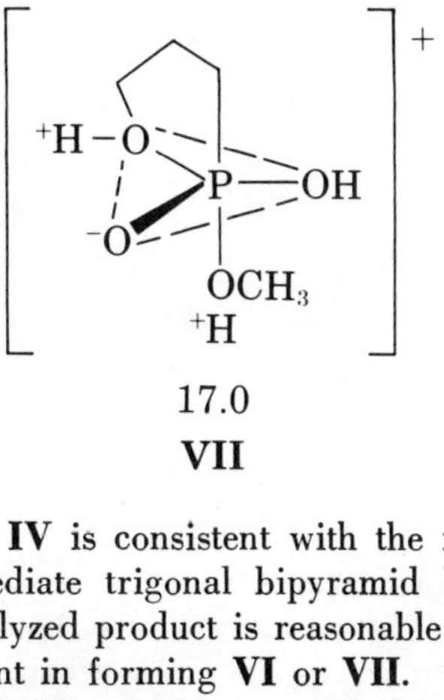

22.0
VI

17.0
VII

Thus, the high rate of hydrolysis of **IV** is consistent with the relief of ring strain on formation of the intermediate trigonal bipyramid **V**. The retention of the OCH_3 group in the hydrolyzed product is reasonable in view of the high barriers suspected to be present in forming **VI** or **VII**.

Ethyl Ester of Butylphosphinic Acid. Consistent with this type of mechanism is the slow hydrolysis rate found for cyclic five-membered phosphinates. (*17, 19*). The rate of hydrolysis (*17, 19*) of the ethyl ester of the cyclic phosphinic acid **VIII**, proceeding with apparent ring retention (*17*), is comparable with that of the corresponding ester of diethylphosphinic acid (Table 2.2). Neither ring opening nor hydrolysis external to the ring leads to a low energy path. In the former, nucleophilic attack by a water molecule leads to the intermediate state, **IX**. However, an alkyl group resides in an axial position contrary to the electronegativity rule. Similarly, for hydrolysis external to the ring, pseudorotation of **IX** is necessary to bring an alkoxy group into a leaving position and leads to the high energy structures **X** and **XI**.

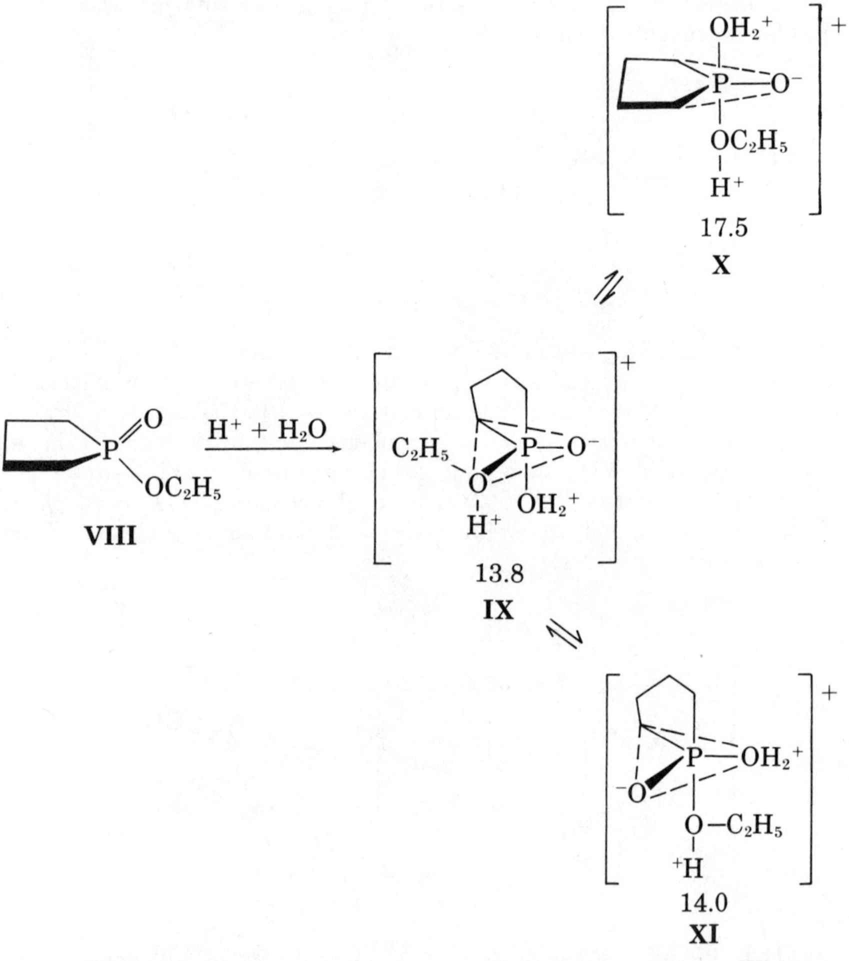

Approximately an 8–13 kcal/mol increase in ring strain has been calculated (27) to accompany a change in the O–P–O angle of a five-membered phosphorus-containing ring from 90° to 120°. Further NMR data (30) on model cyclic oxyphosphoranes (Volume I, p 186), whose ligands bear a close correspondence in electronegativity with those for the intermediate states under discussion, show a relatively large free energy of activation for intramolecular ligand exchange. Barriers for the latter process are presumed (30) to be associated with the placing of a ring alkyl group in an axial position (10–17 kcal/mol) or in formation of a diequatorial ring (20 kcal/mol) in trigonal bipyramidal intermediates. The inference here is that **X** may be somewhat less stable than **XI**. Additional support for this comes from an examination of the square pyramidal states encountered in the formation of **X** and **XI** from **IX**. These are **Xa** and **XIa**, respectively. The rules set forth at the beginning of this chapter governing square pyramidal stability indicate that

XIa is of lower energy than **Xa**. Calculation via the numerical scales substantiates this contention.

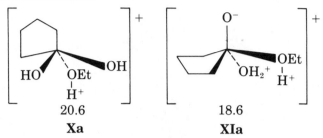

20.6   18.6

Xa   **XIa**

Accordingly, elimination of an ethanol molecule is expected to complete the hydrolysis process via **XI**.

To account for the comparable rates of hydrolysis between the cyclic phosphinate **VIII** and its acyclic analog, one must argue that the energy gained in releasing ring strain (formation of **IX**) is largely counterbalanced by the loss in energy associated with the resultant placement of an alkyl group in an axial position. On this basis, it is concluded that the energy associated with the transition state leading to the formation of **IX** is an effective representation of the barrier to hydrolysis.

A potential energy diagram of the following form is visualized, where the latter transition state is labeled T_1 (*see* p 91). Both **IX** and **XI** are

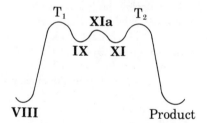

similarly structured and have groups of comparable electronegativities. The energies are undoubtedly close, and interconversion between them is probable, at least at a rate comparable with the rate of hydrolysis. As a result of the latter interpretation, a relatively flat potential surface is expected between **IX**, **XI**, and the transitory square pyramid **XIa**. Thus, both the slow rate (formation of **IX**) and the fact that the acid hydrolysis apparently proceeds with ring retention (rapid formation of **XI**) is largely explained.

Of course, **X** could be formed directly from **VIII** without the intervention of pseudorotation. Its energy estimated at 17.5 kcal/mol is comparable with that estimated for **XIa**—i.e., the barrier between **IX** and **XI**. Thus, this pathway is also consistent with ring retention for **VIII** and a hydrolysis rate which is comparable with its acyclic analog.

Acyclic phosphinates hydrolyze slowly with inversion of configuration (*31*). In terms of a trigonal-bipyramidal representation, the most favorable intermediate seems to be **XII** analogous to **X** for a related cyclic derivative.

This formulation is in accord with the suggestion of Green and Hudson (*31*) and is also that expected on the basis of substituent electronegativities.

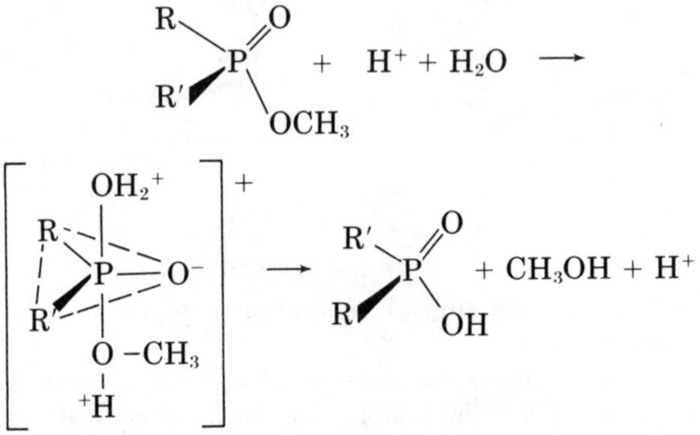

Extended Hückel Treatment. Boyd (*32*) carried out a molecular orbital treatment of cyclic and acyclic phosphate esters via the extended Hückel theory. His results support the main arguments of the hydrolysis mechanism advanced by Westheimer and co-workers (*1, 16*). Specifically, the least energy pathway for nucleophilic attack on the phosphate leads to a trigonal bipyramid with the entering and leaving groups occupying axial positions. For the hydrolysis of methylethylene phosphate, pseudorotation through structures analogous to **I**, **II**, and **III** gives a minimum potential surface, i.e., within the limitations imposed by the Hückel theory and the range of structures investigated. An upper limit of 12–15 kcal/mol resulted for the barrier to pseudorotation, corresponding to the energy diffeernce between **XIV** minus **XIII**.

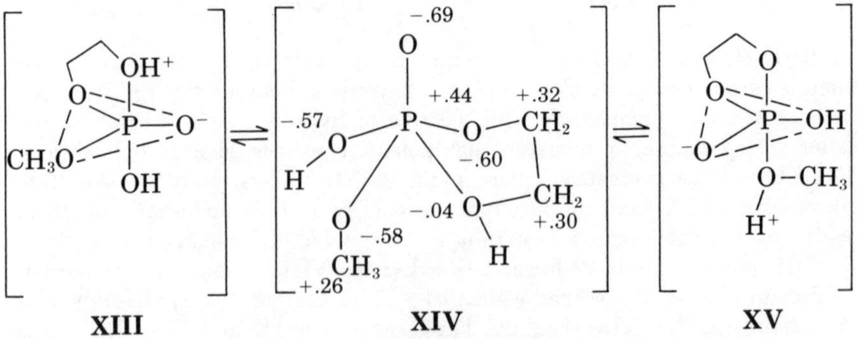

(charge densities from Ref. 32)

Considering the approximate nature of the calculation for such a complicated system, the magnitude of the barrier energy should not be taken too seriously. As Boyd (*32*) points out, the barrier would most likely be lower if optimal geometries were used. The numerical estimate from application of the substituent scales—1.9 kcal/mol—is in a more acceptable range.

Oxyphosphorane Ester Rate Comparisons. It is worthwhile to compare substituent exchange behavior observed for oxyphosphorane compounds (Volume I, p 177) that closely resemble intermediates postulated in the phosphate ester hydrolysis mechanisms. Adduct **XVI**, formed from biacetyl and trimethylphosphite (*33*), is comparable in proposed structure (*34*) and ligand electronegativity to the principal Intermediates **I** and **III** formed in the acid hydrolysis of methylethylene phosphate (p 69). The rapid rate in the latter hydrolysis, postulated to proceed via pseudorotation (**I → III**) parallels the NMR observation (*35, 36*) that intramolecular ligand exchange (pseudorotation) for the related oxyphosphorane **XVI** is rapid even at −100°C.

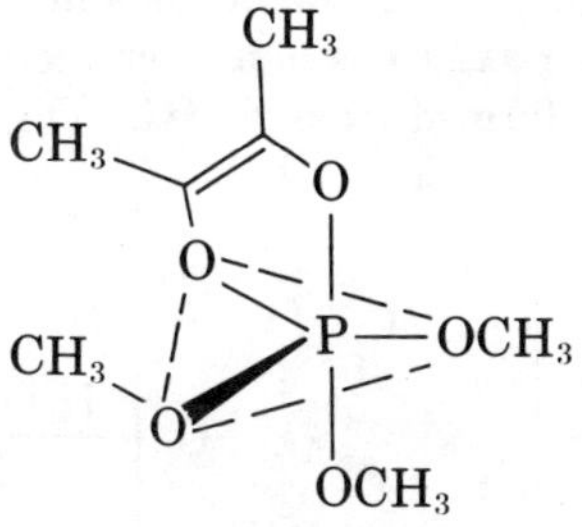

XVI
ligand exchange,
rapid at −100°C

Adduct **XVII**, formed from benzylidene acetylacetone and trimethylphosphite, appears stereochemically more rigid since intramolecular ligand exchange is stopped at −65°C as shown by temperature-dependent NMR data (*1, 34, 37*). This behavior is consistent with the rapid hydrolysis of the methyl ester of propylphostonic acid **IV** proceeding through the intermediate **V** but not involving pseudorotation. The lower exchange rate observed for structures of the type **XVII** suggests that the rate of the pseudorotational

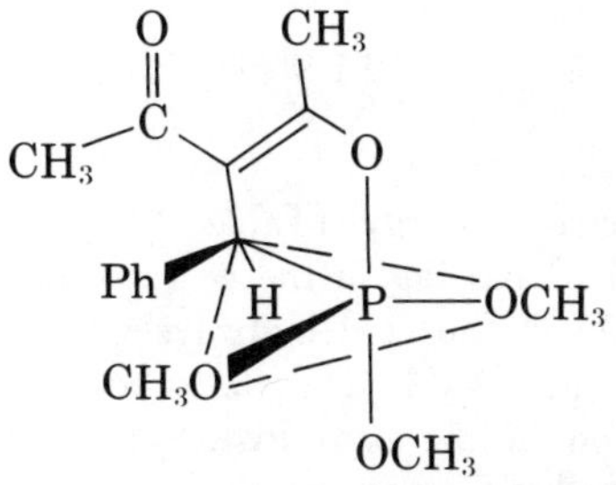

XVII
ligand exchange,
stopped at −65°C

process of **V** leading to intermediate states similar to **VI** or **VII** is sufficiently reduced compared with overall hydrolysis kinetics to be of little significance in determining product formation.

Oxyphosphoranes that are related structurally to Intermediate **IX** formed in the slow hydrolysis of the cyclic phosphinic acid **VIII** do not seem to be available for comparison.

Alkaline Hydrolysis of Cyclic Esters

In contrast to the acid hydrolysis of methylethylene phosphate that results in approximately 30–50% ring retention, alkaline hydrolysis (*1, 24*) at pH 10–13 proceeds with nearly 100% ring opening. If ring retention were to occur, the group departing in the alkaline hydrolysis would be a methoxy group present in an axial position after the required pseudorotation of the intermediate initially formed from **XVIII**. This group is apparently of

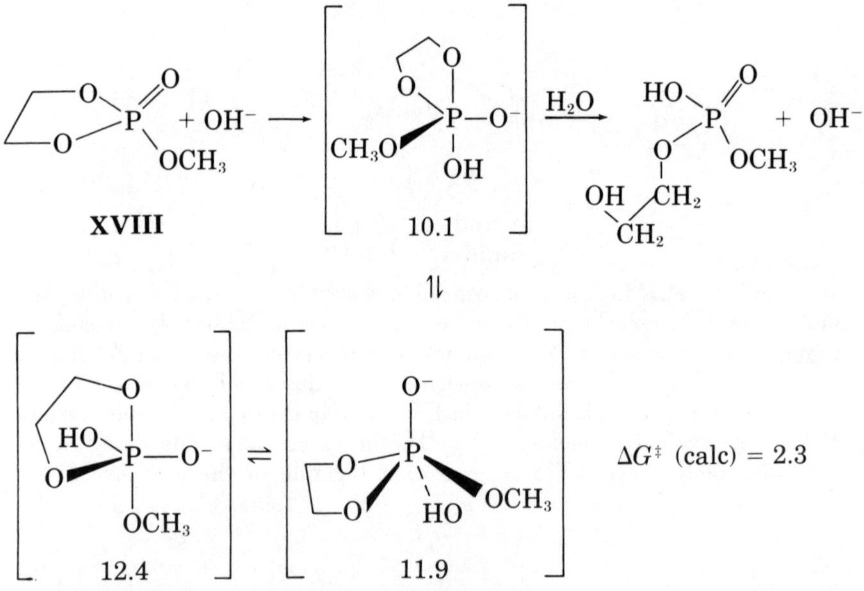

$$\Delta G^{\ddagger} \text{ (calc)} = 2.3$$

sufficiently reduced electronegativity (Table 2.1) compared with the protonated form (CH_3OH^+), departing in the corresponding pseudorotated conformer (**III**) encountered in the acid hydrolysis that pseudorotation is not significantly competitive for **XVIII**. The numerical estimate of the activation energy for pseudorotation in the base hydrolysis (nearly double that calculated for the acid hydrolysis) supports this contention. The main driving force for the reaction, however, is still present—the relief of ring strain on formation of the intermediate trigonal bipyramid. Consequently, the rapid base hydrolysis proceeding with ring opening comparable with the rate for acid hydrolysis (Table 2.2) is explained reasonably.

In very strong alkali a significant amount of cyclized product is again obtained. Westheimer and co-workers postulate (24) that this medium should favor the formation of dianionic species as a result of the ionization of both hydroxyl groups in the pentacoordinated intermediate. Since these anions prefer equatorial positions, pseudorotation is facilitated and exocyclic cleavage is enhanced. This agrees with our numerical estimates, showing that the pseudorotated conformer **XX** is lower in energy than **XIX** and is formed with no intervening barrier.

No new mechanistic features are needed to account for the rapid alkaline hydrolysis observed (12, 17) for cyclic phosphonates similar to **IV** ($\sim 10^6$ times more rapid than its acyclic counterpart) or the slow base-catalyzed hydrolysis of cyclic phosphinates (17, 19) similar to **VIII** (Entry **5**, Table 2.2). In the former, relatively high barriers are estimated for pseudorotation, consistent with the lack of ring retention in the product. The argument followed here is the same as that discussed for the acid hydrolysis of **IV**. The numerical estimate shows that the barrier to pseudorotation for the base-catalyzed hydrolysis is higher than that for the acid hydrolysis.

For the phosphinates, alkaline hydrolysis similar to acid hydrolysis (p 96) leads initially to a high-energy intermediate with a ring carbon atom in an axial orientation. A slightly higher pseudorotational barrier is estimated because of the presence of nonprotonated substituents of reduced electronegativity (OH and OCH₃). However, because of the overall inertness to hydrolysis of this class of compounds, the rate of pseudorotation need not be too great to make a significant contribution to the kinetic scheme. The extent of ring retention does not seem to have been reported for the base hydrolysis of phosphinates. Thus, assessment of the detailed reaction path must be deferred.

Rate data on other cyclic and acyclic phosphorus esters measured by Aksnes and Bergesen (17) are summarized in Table 2.3. The rates can be interpreted in terms of the discussion above. However, the large rate acceleration observed for the Phosphonate **A** compared with that for the larger six- and seven-membered cyclic Structures **B** and **C** as well as the acyclic Analog **G** is partly a result of a lower activation energy and partly a result of a favorable entropy factor. The former may reflect a weaker P–O bond (17)

Table 2.3. Rate Constants and Activation Parameters
Phosphinates, Phosphonates,

Com-pound	Solvent	Rate Constant: $(L\ mol^{-1}\ sec^{-1} \times 10^4)$		
		$30°C$	$40°C$	$50°C$
A	Water	19250 (0°C)	51250 (13°C)	5.4×10^5 (calcd)
B	Water	—	13.6	26.2
C	Water	—	—	1.096
D	50% Alcohol–water	—	—	1.18
E	80% Alcohol–water	—	—	0.300
F	50% Alcohol–water	—	—	0.730
G	Water	0.407	0.872	1.72
H	Water	—	—	$\sim 10^7$
I	Water	—	18.1	36.0
J	Water	12.8 (0.1°C)	48.9 (15°C)	74.1 (20°C)
K	Water	—	0.249	0.520

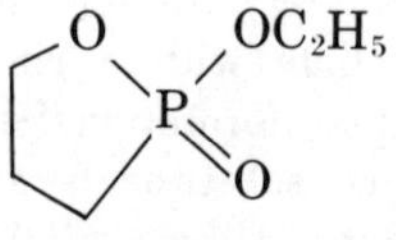

A

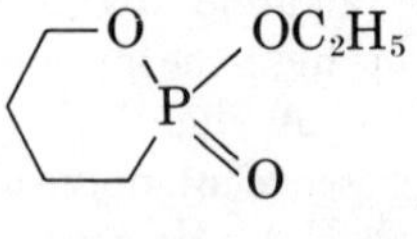

B

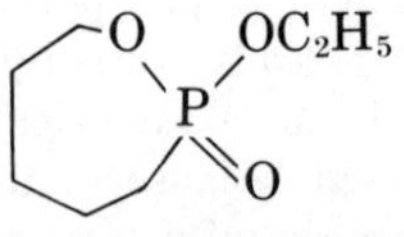

C

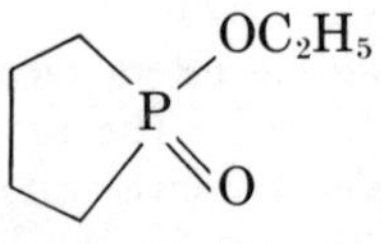

D

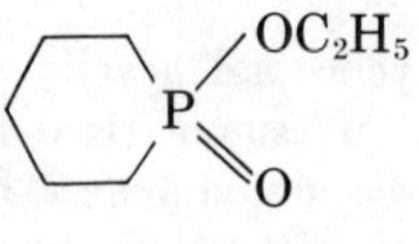

E

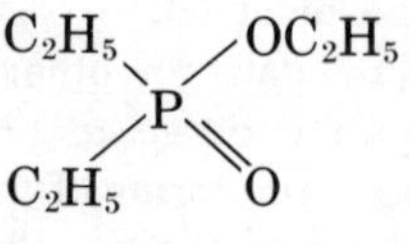

F

for the Alkaline Hydrolysis of Cyclic and Open Chain and Phosphates (*17*)

Rate Constant ($L\ mol^{-1}\ sec^{-1} \times 10^4$)		$\Delta E^{\neq}$ (kcal/mol)	Log A	$\Delta S^{\neq}$
60°C	70°C			
—	—	11.7 ± 1	9.64	−16.6
50.3	—	13.7 ± 0.5	6.65	−30.3
2.109	4.056	14.0 ± 0.2	5.46	−35.7
2.24	4.22	14.1 ± 0.5	5.60	−35.0
0.568	1.08	14.2 ± 0.5	5.17	−37.0
1.39	2.63	14.3 ± 0.5	5.53	−35.2
—	—	14.0 ± 0.5	5.76	−34.2
—	—	—	—	—
68.7	125	13.8	6.87	−29.3
163 (30°C)	—	14.0	8.30	−22.5
1.042	2.00	14.8	5.75	−34.4

Acta Chemica Scandanavica

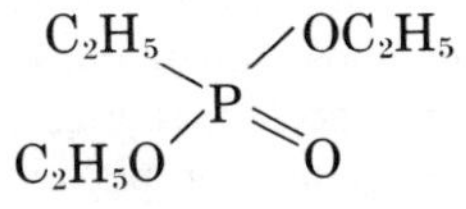

G 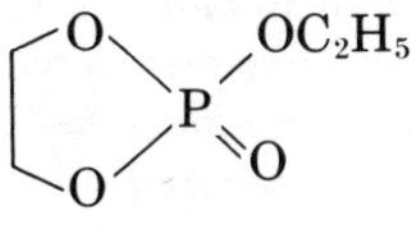 H 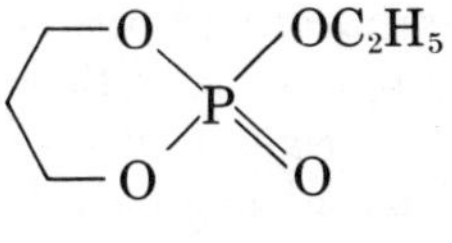 I

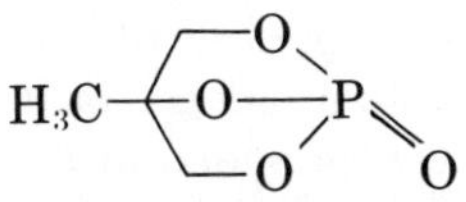

J 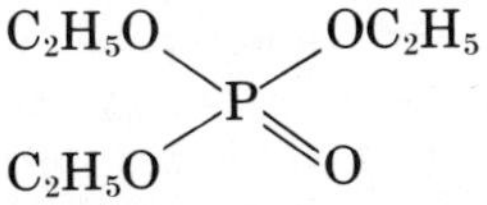 K

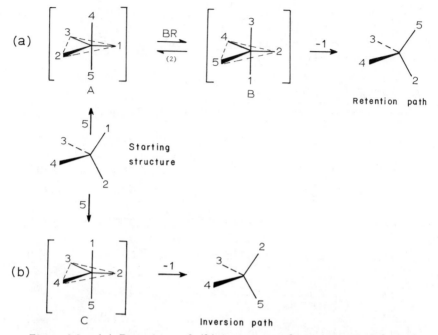

Figure 2.1. (a) Retention and (b) inversion paths at a tetracoordinated chiral center proceeding via trigonal bipyramidal intermediates. If a ring structure is present, cyclization at Positions 3 and 4 is assumed. For the intermediate undergoing pseudorotation, the number in parentheses refers to the apical atom in the square pyramid that is traversed in the BR process— i.e., the pivotal ligand in the Berry process.

in the five-membered pholane ring compared with that in the others in agreement with the ring-strain hypothesis (12, 23). The relative increase of entropy of activation for the five-membered cyclic phosphonate may reflect a less rigid pentacoordinate intermediate (38) with accompanying greater degree of motional freedom compared with intermediates for phosphonates containing larger rings or no rings.

In this context, Westheimer (1) pointed out that the rate difference in base hydrolysis between methylethylene phosphate and acyclic trimethylphosphate (Table 2.2) corresponds to a difference in $\Delta G^{\ddagger}$ of ca. 8.5 kcal/mol (or $\Delta E^{\ddagger}$ of ca. 7.5 kcal/mol). However, the difference in heats of hydrolyses is only 5–6 kcal/mol. If the latter energy reflects strain energy in the five-membered cyclic structure that is released during the formation of the Intermediate **I**, it could account for a large part (1) but not all of the explanation for the high rate of reaction of this cyclic ester. The particular form that stabilizing entropic effects might take have been discussed by Gorenstein et al. (28).

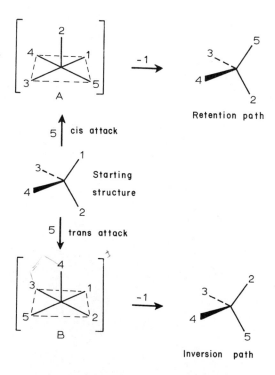

Retention path

Starting structure

cis attack

trans attack

Inversion path

Figure 2.2. Retention and inversion paths at a tetracoordinated chiral center proceeding via square pyramidal intermediates. If a ring structure is present, cyclization at Positions 3 and 4 is assumed.

General Reaction Paths

The mechanisms just discussed were necessarily detailed to illustrate the potential usefulness of the application of principles of pentacoordination to mechanistic considerations. Most other phosphorus reactions that are postulated to proceed via pentacoordinated intermediates are amenable to a similar approach. As we have seen, there are two major pathways: nucleophilic attack (typical of S_N2 displacements for four-coordinate carbon chemistry), leading to inversion of configuration, and nucleophilic attack with accompanying pseudorotation, leading to retention of configuration. This general scheme, which represents the one most widely postulated, is outlined in Figure 2.1.

For our purposes, we define **A, B,** and **C** as intermediates (p 90). A possible potential energy profile for the retention route (Figure 2.1a) has been outlined in connection with the acid hydrolysis of the cyclic phosphinate, **VIII** (p 97). Relative to the latter diagram, Intermediates **A** and **B** of

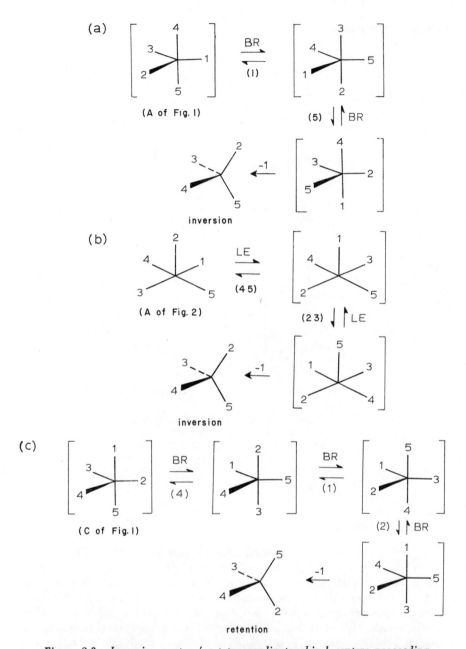

Figure 2.3. Inversion routes for tetracoordinate chiral centers proceeding via (a) a trigonal bipyramidal intermediate and (b) a square pyramidal intermediate. In (c) a retention route is represented proceeding via a trigonal bypyramidal intermediate. If a ring structure is present, cyclization at Positions 3 and 4 is assumed. However, Process c applies only to acyclic derivates. See Figure 2.1 for further definitions. LE refers to ligand exchange between square pyramidal intermediates. The accompanying number pair in parentheses refers to the axial positions of the trigonal bypyramid traversed in the ligand exchange process.

Figure 2.1a are equated to **IX** and **XI**, respectively, and the square pyramidal transition state formed during the pseudorotation between **A** and **B** replaces **XIa**.

T_1 and T_2 represent transition states (*see* p 91) leading to the formation of intermediates (**A** and **B**) and normally provide the rate-limiting features of the reaction. It is assumed that a correspondence exists between the energies of the transition states T_1 and T_2 and their respective intermediates, **A** and **B**. If the pseudorotational barrier is high relative to the rate of the reaction, the competitive process through an intermediate of the Type **C** of Figure 2.1b takes over. By using chiral phosphorus centers, these two fundamental processes can be differentiated for specific reactions. Much work has been done in this area, both with acyclic and cyclic substituents. Some of these topics are treated later in this chapter, but first we explore some mechanistic alternatives.

Although not normally postulated, a complementary scheme (Figure 2.2) can be written for square pyramidal intermediates that are formed as a result of nucleophilic attack at tetracoordinated phosphorus. We assume that groups enter and leave from basal positions of the square pyramid since these positions are predicted to involve the longer and weaker bonds for an idealized square pyramid (Volume I, p 116). Pseudorotation is not a required process for either retention or inversion to take place via square pyramidal intermediates.

When small-membered rings are present, however, both schemes (Figures 2.1 and 2.2) place the ring in an unfavorable orientation for the reaction to proceed by an inversion mechanism, i.e. diequatorial placement in the trigonal bipyramidal Intermediate **C** (Figure 2.1), and apical–basal occupancy in the square pyramid, **B** (Figure 2.2). These higher-energy ring configurations may be avoided by requiring additional pseudorotations that maintain the ring in the more favorable axial–equatorial orientation of the trigonal bipyramid or cis dibasal positions of the square pyramid. This can be accomplished with the initially formed intermediates (**A**) shown for each of the retention routes in Figures 2.1 and 2.2. Successive pseudorotation starting with **A** in each case will lead to products with inversion of configuration. These possibilities are shown in Figures 2.3 and 2.3b.

However, steric or electronic factors may more than offset the energy gain as a result of the lesser ring strain in the latter pathways and thus render the inversion routes shown in Figure 2.3 less favorable compared with those illustrated in Figures 2.1 and 2.2.

For acyclic derivatives, it is also possible to form a product with retention of configuration by executing multiple pseudorotations on Intermediate **C** (Figure 2.1). This path is outlined in Figure 2.3c. The decision regarding the operation of multistep processes of the type shown in Figure 2.3, in preference to or in competition with the simpler schemes first outlined, will be determined by the makeup of the particular derivatives, their substituent properties, and whether kinetic or thermodynamic control is more important in guiding the construction of intermediate formulations.

The schemes illustrating square pyramidal intermediates have limited

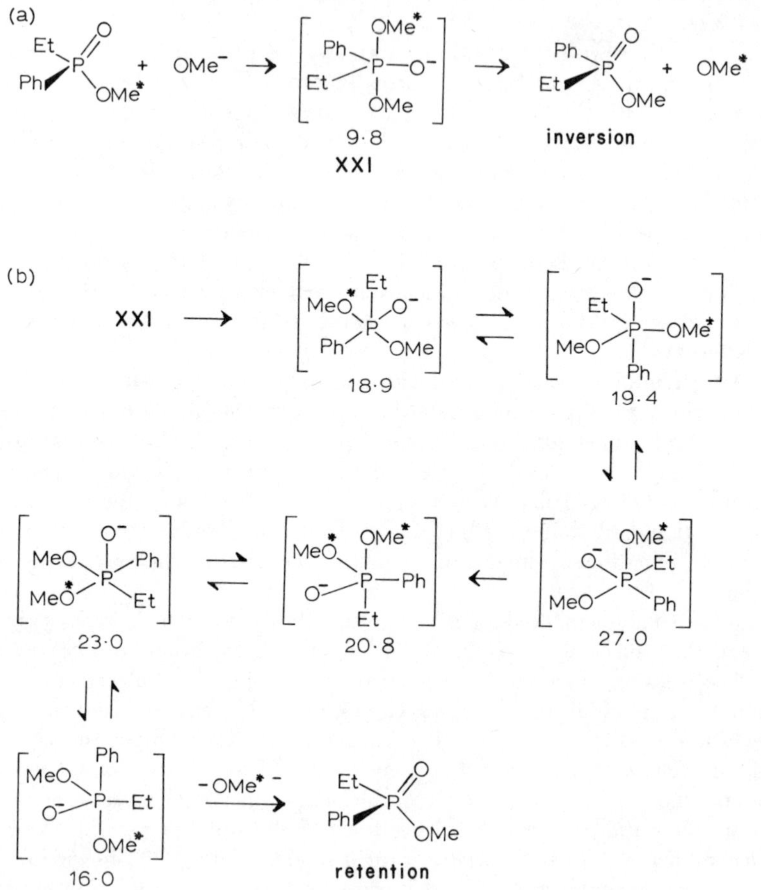

Figure 2.4. Numerical estimates of the energy of intermediates in inversion and retention routes for the racemization of optically active methylphenylethylphosphonite with excess methoxide ion. The route shown in (a) cor-

applicability, confined principally to reactions of spirocyclic derivatives, since only for these derivatives have corresponding phosphorane structures been shown to be nearly square or rectangle pyramidal (Volume I, p 11). Even so, this class of reactions is relatively unexplored, and no decision regarding a detailed mechanism has been forthcoming (p 163).

An example that illustrates the probable importance of the various pathways involving trigonal bipyramidal intermediates is suggested by observations on acyclic phosphinates. These derivatives have been shown to hydrolyze slowly with inversion of configuration (*38*). Green and Hudson (*31*) found that the rate of racemization of optically active methylphenylethyl phosphinate *responds to the generalized scheme in Figure 2.1b; (b) corresponds to the scheme in Figure 2.3c; (c) corresponds to the scheme in Figure 2.3a; (d) corresponds to the scheme in Figure 2.1a. The favored Route a agrees with the observed inversion of configuration (31).*

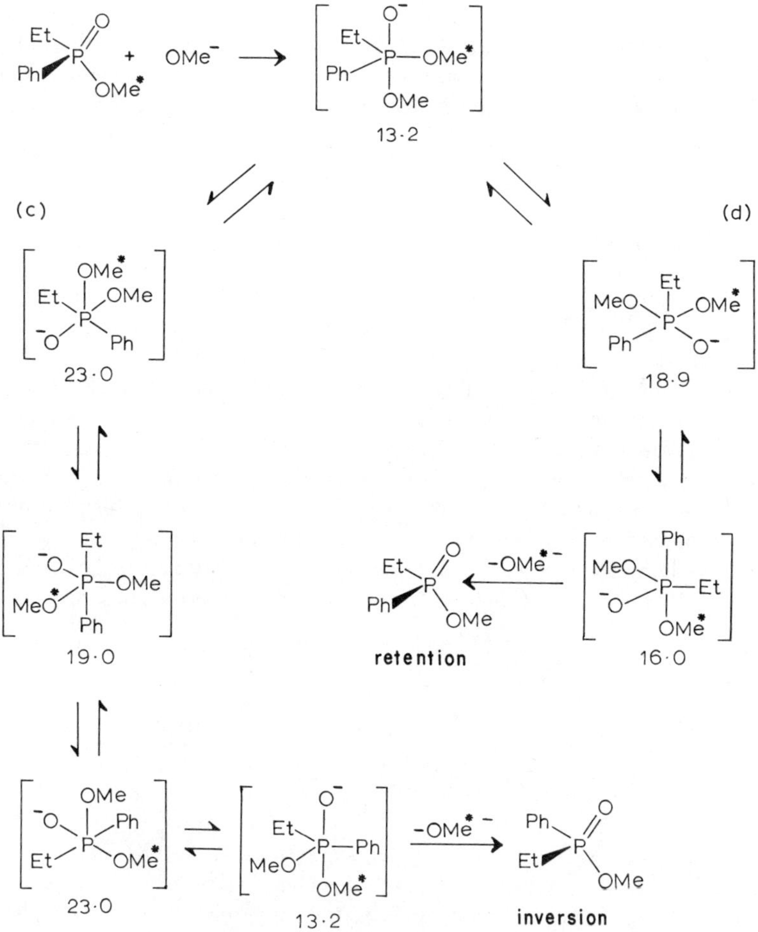

by excess methoxide ions equaled twice the rate of isotopic exchange of ^{14}C-labeled methoxy in the phosphinate, thus proving that the reaction is stereospecific. Figure 2.4 shows possible intermediates arising in the two inversion processes and in the two retention pathways illustrated in Figures 2.1 and 2.3 for trigonal bipyramidal intermediates. The low-energy process, based on numerical estimates, is the inversion path via **XXI**. All other paths are estimated to be considerably higher in energy, with the least-favored process being the retention pathway requiring three pseudorotations starting with **XXI**.

Rate Enhancement Effects in Other Systems

Hydrolysis of OS–Ethylene Phosphorothioate. Gay and Hamer
(*39, 40*) studied the pH rate-product profile for the aqueous hydrolysis of
OS–ethylene phosphorothioate (**XXII**, R = OCH₃). They found that the rate

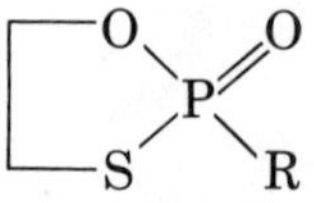

$$R = OCH_3, O^-, Ph$$

XXII

dependence was similar to that obtained for the hydrolysis of methylethylene
phosphate (p 93). In contrast to the latter study, however, no ring retention
was observed. Although endocyclic P–O bond cleavage predominates in acid
solution below pH of 1.5, almost exclusive P–S bond cleavage results in
neutral or alkaline solution. Further, the hydroxide rate constant for the
cyclic thioate is much larger than that for methylethylene phosphate. The
latter rate acceleration is typical. For example, similar rate accelerations are
found (*41*) for the alkaline hydrolysis of OOS–triesters of phosphorothioric
acid compared with those of analogous trialkyl phosphates.

The dominance of P–O bond cleavage in strongly acid solution is consistent with the expected increased concentration of protonated species, like

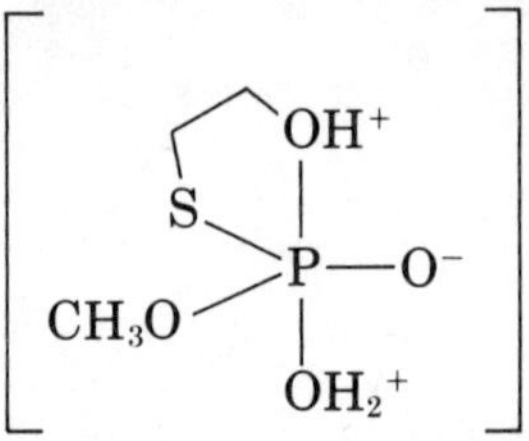

that postulated by Westheimer and co-workers (*24*) (p 93) in the case of
the acid hydrolysis of methylethylene phosphate below pH 2. In agreement
with the electronegativity rule (p 88), the ring oxygen atom occupies the
axial site in the intermediate trigonal bipyramid. Pseudorotation is required
to bring the ring sulfur atom into a leaving position. However, as discussed
for the hydrolysis of methylethylene phosphate, pseudorotation becomes less

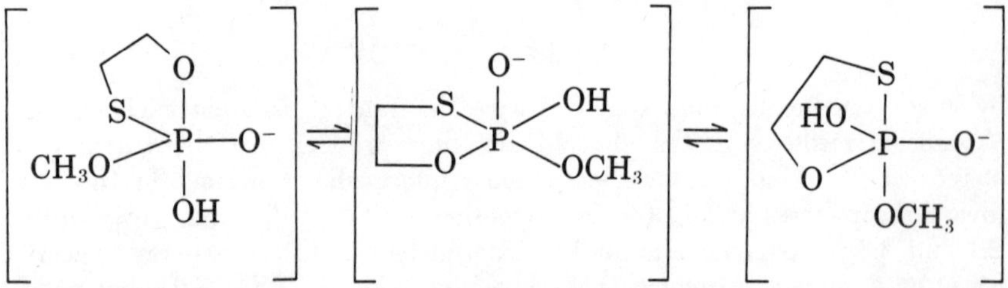

probable as solution basicity increases. This is even more true for the sulfur
derivative because of the introduction of an atom of reduced electronegativity.

Possibly the weakness of the P–S linkage acts as a driving force for the pseudorotational process and accounts for the enhanced rates observed, with thio substituents undergoing P–S bond cleavage. Estimates (*42*) of sigma bond strengths, which include ionic contributions (*43*), suggest 55 kcal/mol for the P–S bond, much lower than the 84 kcal/mol for the strength of the corresponding P–O bond. Thus, pseudorotation leading to the placement of the ring P–S bond in the weaker axial site of the trigonal bipyramid should result in a "loosening" of the structure, which facilitates P–S bond cleavage over release of the methoxy group at the opposite axial site. The process might be particularly favored in basic solution as anionic charge increases. The accompanying rise in ligand repulsion should contribute to the rupture of the P–S linkage. Evidence, to be discussed later (p 129), indicates that alkythio ligands are better leaving groups than alkoxy ligands in alkaline solution and lends support to the hypothesis advanced here.

Replacement of the methoxy group of the cyclic thioate (**XXII**) with the anion O^- or the phenyl group (the latter giving the phosphonate analog) results in base solvolysis proceeding with endocyclic P–O bond cleavage (*39*). The use of these less-electronegative substituents is expected to increase the pseudorotational barrier since they would occupy axial positions in the respective pseudorotated intermediates. Accordingly, endocyclic cleavage of the P–S bond becomes competitively less attractive compared with direct S_N2 displacement (Figure 2.1b).

Highly Strained Cyclic Phosphinates. In contrast to the slow rates of hydrolysis found for cyclic five-membered phosphinates (pp 92 and 102), bicyclic phosphinates such as shown undergo hydrolysis of the ester group

located at the bridge phosphorus atom (P_1) much more rapidly than the ester group attached to the second phosphorus atom (P_2) in either acid or base (*1, 44, 45, 46*). The hydrolysis of the ester group at P_2 is comparable with that of monocyclic five-membered phosphinates or their acyclic analogs.

Presumably, a high degree of strain energy is present in these bicyclic derivatives. Formation of a trigonal-bipyramidal intermediate allows relief of the strain with the energy gain exceeding the barrier to placing a carbon atom in an axial position (*44, 45, 46*). Consequently, a low-energy pathway seems feasible, proceeding through an intermediate of the type:

Table 2.4. Alkaline Hydrolysis of Phosphinate and Phosphinite Esters[a]. Second-Order Rate Constants[b] (47)

$$\overset{O}{\underset{\|}{RP(OC_2H_5)_2}}$$

R	Ethyl	Isopropyl	tert-Butyl
	300/70°C	440/120°C	60/120°C
	1200/90°C		

$$\overset{O}{\underset{\|}{R_2P(OC_2H_5)}}$$

R	Ethyl	Isopropyl (1N)	tert-Butyl (2N)
	260/70°C	10/100°C	0.08/120°C
		41/120°C	

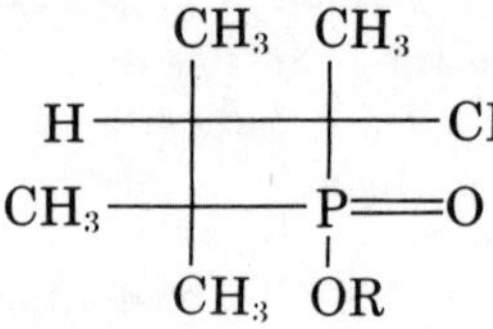

	R = Ethyl	R = Methyl	**XXIII**
	300/70°C	840/70°C	390,000[c]/45°C

[a] Hydrolyses were carried out under pseudo-first-order conditions in 0.1N NaOH except where noted.
[b] Lmol⁻¹ sec⁻¹ × 10⁶.
[c] pH = 10.0.

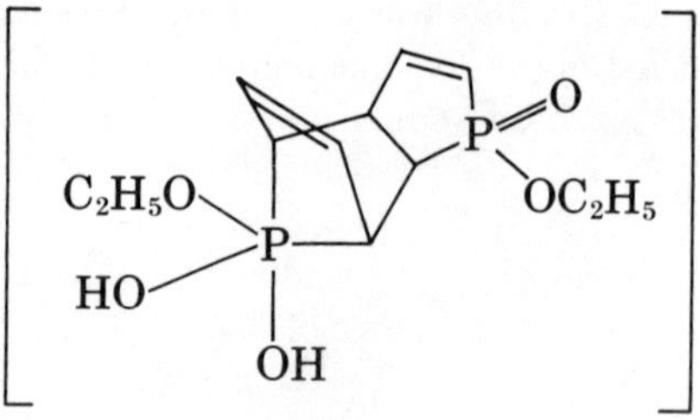

15.1
Initial Intermediate **A**

Pseudorotation of the latter (it was shown that no structural reorganization accompanies hydrolysis (44, 45, 46)) brings the ethoxy ligand to a departing axial site:

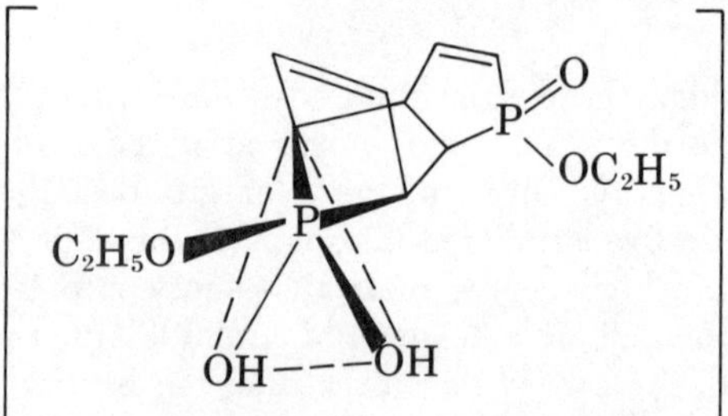

Pseudorotated Form **B**

Kluger and Westheimer (*44*) comment that perhaps in cases like this, the tetragonal pyramid (**C**) formed during pseudorotation between **A** and **B** is actually an intermediate and that the trigonal bipyramid (**A**) may be a transition state (*see* p 91). Although the substituent scales we discuss

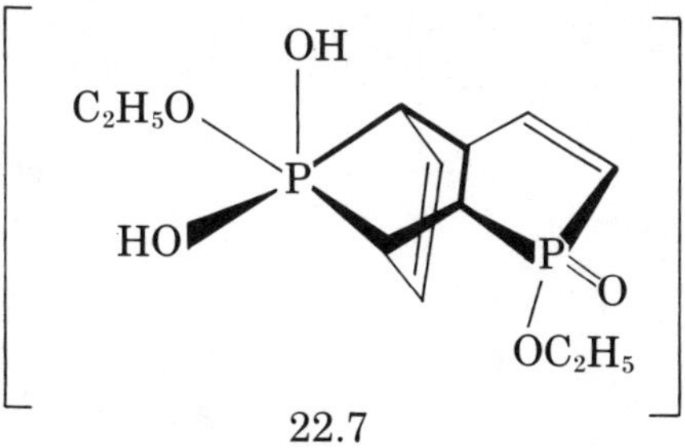

22.7

C

(p 89) have not been properly calibrated for highly strained systems such as these, their application shows considerable preference for the trigonal-bipyramidal Intermediate **A** (15.1 kcal/mol) vs. **C** (22.7 kcal/mol).

Four-Membered Cyclic Phosphinates. Hawes and Trippett (*47, 48*) observed extremely rapid alkaline hydrolysis of the cyclic phosphinate ester, **XXIII** (Table 2.4). The latter contrasts with the slow rates observed for five-membered cyclic phosphinates already discussed (pp 92 and 102). Presumably the increased strain associated with the smaller four-membered ring is

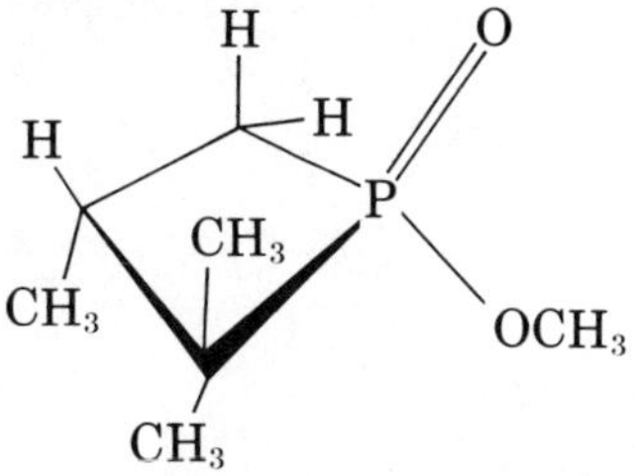

XXIII

largely responsible for the rate acceleration. Formation of the pentacoordinate Intermediate **XXIV** relieves this strain. Thus, additional justification is provided for the formulation of the intermediate states **IX** and **XI** in the hydrolysis of the analogous five-membered cyclic esters (p 96). The energy associated with the relief of ring strain on formation of **XXIV** relative to that in the corresponding five-membered system **IX** must exceed, by a sizable margin, the energy that would be gained if the more electronegative alkoxy group were placed in an axial position and the four-membered ring were diequatorial, **XXV**. (Location of the *gem*-dimethyl groups, as formulated by Hawes and Trippett (*47, 48*) at an axial site in the intermediate is consistent with the operation of kinetic control that appears to take place when a bulky group is present. This effect is discussed in more detail on p 132 for a related phosphinate ester, **XXXIII**, shown on p 119.

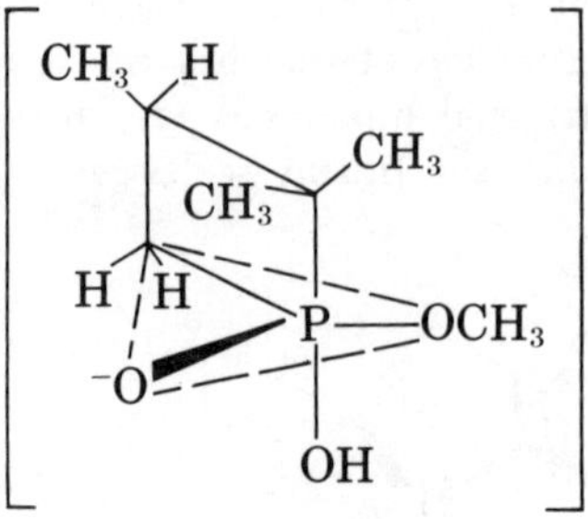

23.4

XXIV

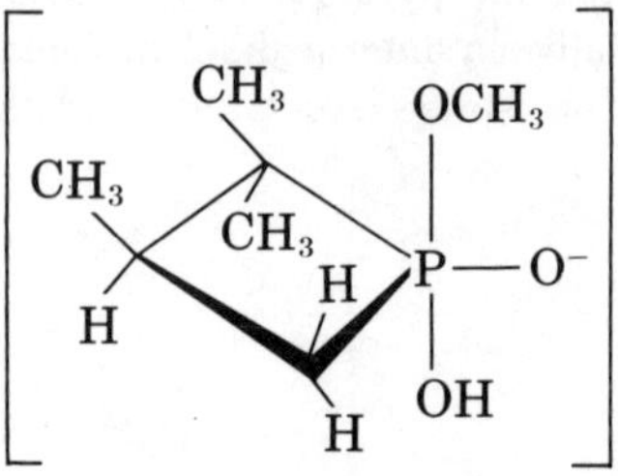

32.4

XXV

Pseudorotation of **XXIV**, pivoting about the equatorial P–O bond, allows the methoxy group to occupy an axial site prior to elimination. A relatively

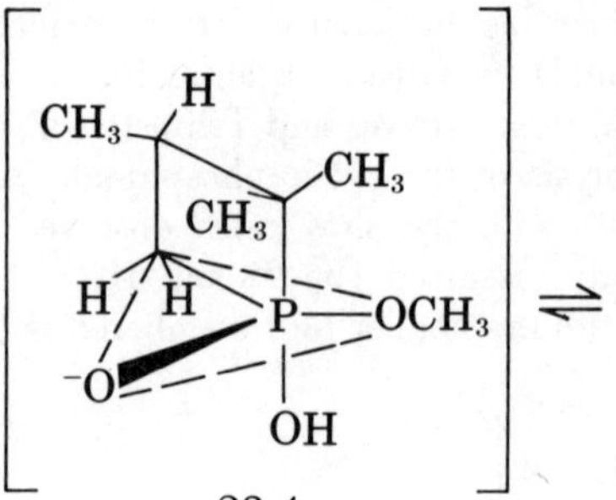

23.4

XXIV

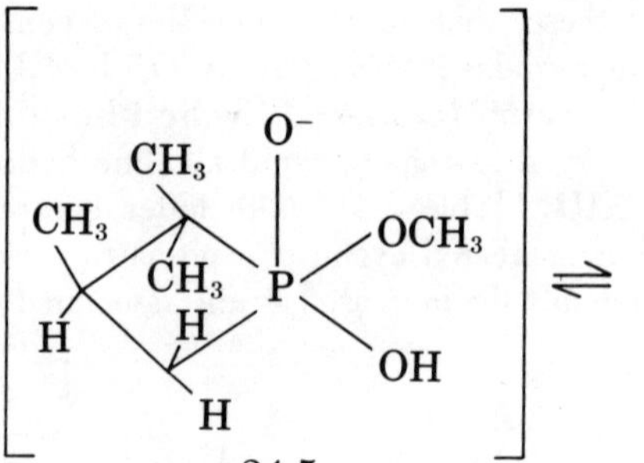

24.5

XXIVa

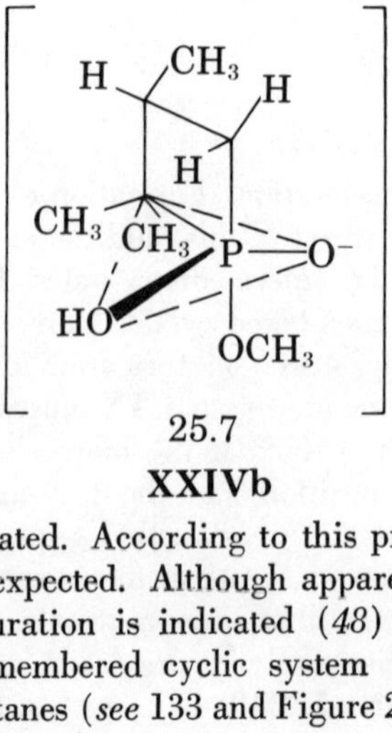

25.7

XXIVb

low barrier is estimated. According to this process, a product with retention of configuration is expected. Although apparently not established for **XXIII**, retention of configuration is indicated (*48*) for related phosphonium salts containing a four-membered cyclic system (p 135) and for more highly substituted phosphetanes (*see* 133 and Figure 2.15, p 148).

The retention pathway, in general, is predicted to be highly favored for nucleophilic attack on tetracoordinated phosphorus compounds containing four-membered ring systems compared with corresponding reactions of five-membered ring systems, primarily because of the greater difference expected in the stability of the initial intermediates proposed for these routes in each case. For the specific members discussed, the estimated activation energies for the "retention" and "inversion" routes for **VIII** (p 96) are somewhat comparable, but for **XXIII** the intermediate along the inversion route (**XXV**) is 7 kcal/mol higher than the highest-energy pentacoordinate form encountered along the retention route. However, if unusual steric effects or groups of poor leaving ability are present in an already-strained ring system, the above generality may not hold (*see* p 138).

Five-Membered Cyclic Phosphonium Salts. Although five-membered cyclic phosphinate esters hydrolyze slowly (*17, 19*), the five-membered cyclic phosphonium salts show a rate enchancement in their hydrolysis reactions. (The same comparison exists between four-membered cyclic phosphonium salts and their neutral esters (p 113) although, as seen in Table 2.4, the four-membered cyclic phosphinate ester (**XXIII**) shows a rate enhancement relative to its acyclic analog; whereas the five-membered cyclic phosphinate ester (Entry **5** in Table 2.2, hydrolyzes at a rate comparable with its acyclic analog.) For example, Aksnes and Bergesen (*49*) showed that cyclotetramethylene methylphenylphosphonium iodide **XXVI** undergoes alkaline hydrolysis about 1300 times faster than the corresponding six-membered ring structure **XXVII** (Table 2.5). The rings are preserved in the products. Similar to the intermediate states postulated in the hydrolysis of the four-membered

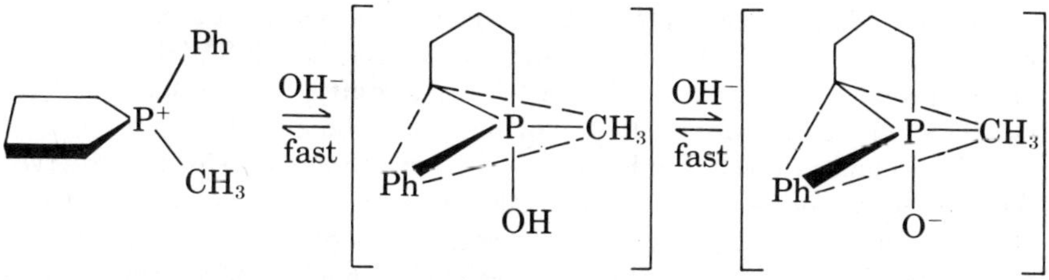

XXVI   XXVIII

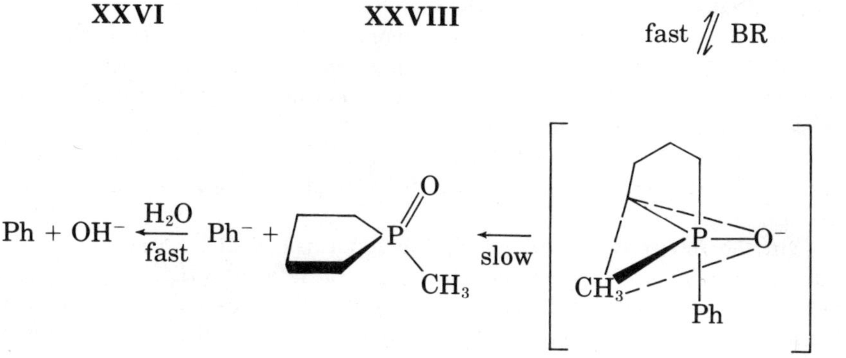

fast ∥ BR

Table 2.5. Rate Data for the Alkaline Decomposition of

Compound	Solvent (wt % ethanol)	Rate constant, ($L^2 mol^{-2} min^{-1}$ at 75°C)
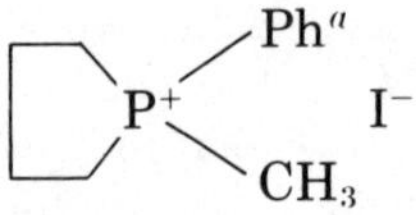 (cyclobutane-ring P^+ with Ph[a], CH_3, I^-)	93.4	1862[c] (calcd)
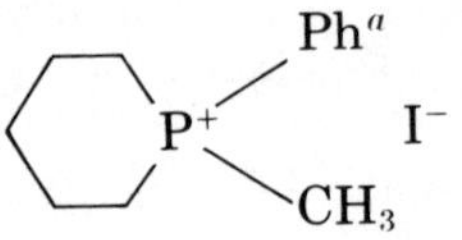 (cyclohexane-ring P^+ with Ph[a], CH_3, I^-)	93.4	1.227
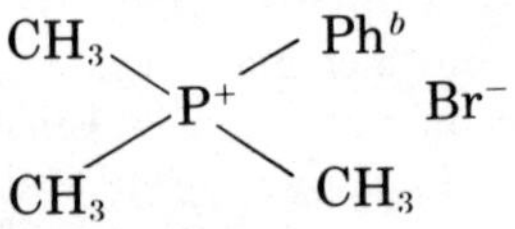 (CH_3, CH_3, P^+ with Ph[b], CH_3, Br^-)	74.8	0.0828
	83.9	0.3105
	93.4	1.673

[a] Ref 49.
[b] Ref 50.
[c] A calculated value based on rate data measured at 35° and 40°C.

cyclic phosphinate ester **XXIII**, relief of ring strain seems to account for the higher rate effect in **XXVI** by formation of Intermediate **XXVIII**. Loss of a phenyl group from an axial position takes place after pseudorotation of the latter and allows ring preservation in the final phosphine oxide.

The complete reaction scheme shown on p 115 agrees with the observed third-order rate dependence, second order in hydroxyl ion and first order in phosphonium ion concentration. In general, the mechanism is similar to that postulated (*51, 52*) for base attack on acyclic phosphonium halides that exhibit identical rate-law dependencies. Although discrete steps are shown, pseudorotation and deprotonation could represent a concerted action. As a result, the electropositive O⁻ ion need not reach an unfavorable axial position along with a ring carbon atom. The details of the mechanism differ substantially from those given by Aksnes and Bergesen. Their discussion (*49*) centers on relief of greater steric effects in **XXVI** compared with **XXVII** that are a result of eclipsing of ring α-hydrogen atoms with hydrogens of the methyl and phenyl groups.

In the relatively unstrained phosphonium salt **XXVII**, base hydrolysis via an inversion pathway (a mechanism typically found (*51, 53, 54, 55*) for acyclic salts) is postulated (*49*). The angle at the phosphorus atom in the hexagonal ring of **XXVII** is expected to be changed little in Intermediate

Phosphonium Salts in Ethanol Water Mixtures (*49*)

$\Delta E^{\neq a}$ (kcal/mol)	$\log A^{d}$
37.6 ($\Delta H^{\neq} = 36.8^{e}$)	26.9 ($\Delta S^{\neq} = 53.6$ eu^{e})
38 ($\Delta H^{\neq} = 37.2^{e}$)	23.7 ($\Delta S^{\neq} = 40.3$ eu^{e})
36	21.7
37	23.1
34.5	21.9

[a] Since the activation data were estimated from rate data at two temperatures 5°C apart, the error is uncertain.
[e] The enthalpies and entropies of activation were estimated from the rate data in Ref *49* by Cremer et al. (*56*). These results seem inconsistent with those in Table 2.10.

Acta Chemica Scandanavica

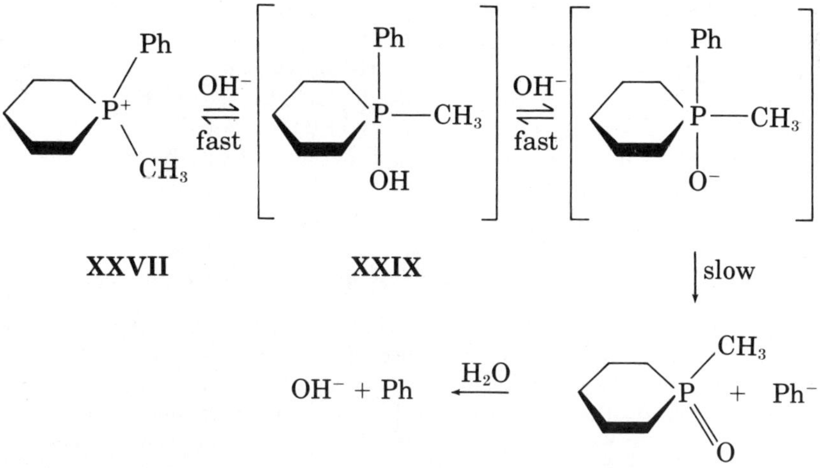

XXIX. Consequently, the rate should approximate that observed in the hydrolysis of acyclic phosphonium halides. In accord with this contention, the rate constant at 75°C for **XXVII** is 1.23 L^{2}mol^{-2}min^{-1}, which compares with 1.67 L^{2}mol^{-2}min^{-1} at the same temperature for the rate constant of the acyclic analog $(CH_3)_3PPh^{+}Br^{-}$, both in 93.4% ethanol (Table 2.5) (*49, 50*).

Rate Influence of an α-Carbonyl Group. The hydrolysis reaction of dimethylphosphoacetoin (**XXX**), an acyclic phosphate containing an α-carbonyl group, has been shown (*57*) to be fully consistent with the mechanistic criteria set forth for five-membered cyclic phosphates (p 92). The presence of the α-carbonyl group greatly enhances the base-catalyzed hydrolysis rate. Ramirez et al. (*58*) observed that Phosphate **XXX** and the corresponding Phosphonate **XXXI** hydrolyzed many orders of magnitude faster than trimethylphosphate and dimethylmethylphosphonate, respectively.

Frank and Usher (*57*) compared the products formed on decomposition of **XXX** and the related Phosphonate **XXXI**. They found that alkaline

$$
\begin{array}{cc}
\underset{\displaystyle CH_3-C-CH-CH_3}{\overset{\displaystyle O}{\overset{\displaystyle \|}{}}} & \underset{\displaystyle CH_3-C-CH-CH_3}{\overset{\displaystyle O}{\overset{\displaystyle \|}{}}} \\
\end{array}
$$

CH$_3$—C—CH—CH$_3$ CH$_3$—C—CH—CH$_3$

O   O

P=O   P=O

CH$_3$O   OCH$_3$ CH$_3$ OCH$_3$

XXX   **XXXI**

hydrolysis of Phosphate **XXX** goes mainly (∼97%) by Path b, whereas hydrolysis of Phosphonate **XXXI** proceeds mainly (∼95%) by Path a. The mechanism suggested allows pseudorotation of the intermediate trigonal bipyramid **XXXII** in the phosphate case but not in that for the phosphonate since an alkyl group would be placed in an axial position reserved for the more electronegative oxygens. The rate for both compounds hydrolyzing by Path a is comparable, the phosphonate hydrolysis being about three times that for the phosphate. Therefore, the high product yield by Path a for the Phosphonate **XXXI** is caused by a decrease in the rate of Path b. Inhibition of pseudorotation thus accounts nicely for the product distribution. Alternative mechanisms (*58, 59, 60, 61, 62*) have been proposed on related systems and should be consulted by the interested reader.

Alkaline Hydrolysis of Acyclic Phosphonium Salts

Up to now we have neglected to mention any orientation effects of attacking groups. The constraints imposed by cyclic systems and substituent electronegativities seemed to be of primary importance in determining intermediate- and transition-state conformations for the reactions already considered. However, rate retardations (with accompanying consequences on the steric course of the reaction) are observed as the bulk of the substituents attached to phosphorus is increased.

Steric Effects. De'Ath and Trippett (*63*) found that alkaline hydrolysis of (−)-benzyl-*tert*-butylmethylphenylphosphonium iodide, **XXXIII**, in 75 vol % ethanol in water proceeds with predominant retention of configuration (>79%). By way of contrast, alkaline hydrolysis of methylethylphenyl-

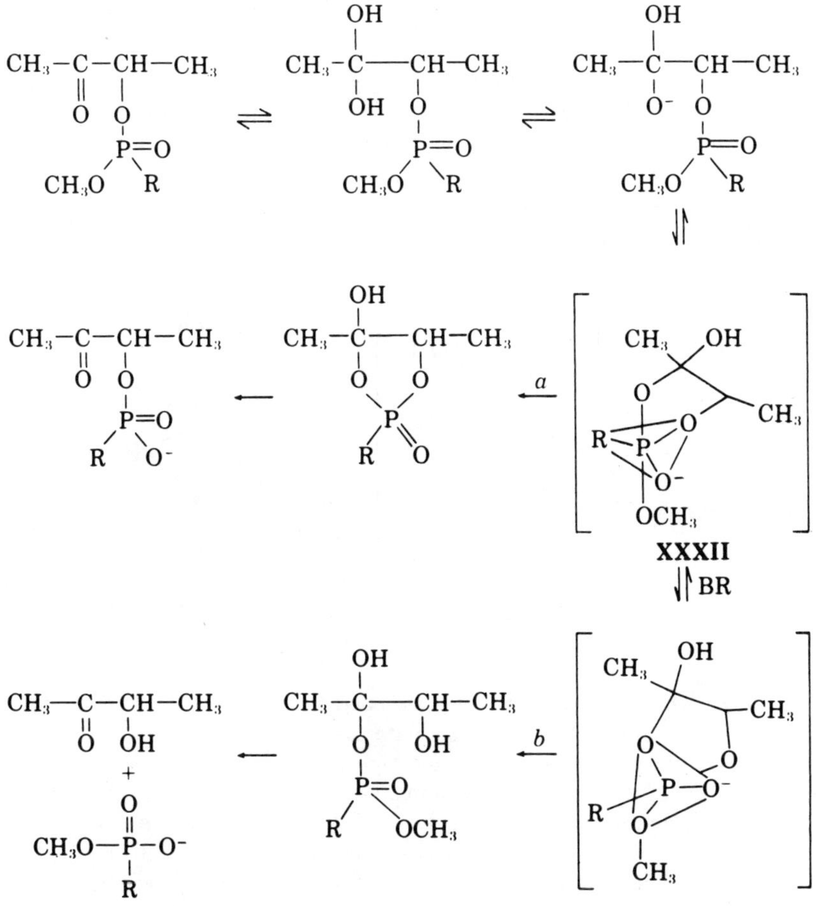

Journal of the American Chemical Society (57)

benzylphosphonium iodide, **XXXIV,** involves complete inversion of configuration (*51, 53, 54, 55*). Here, the group electronegativities are not as vastly different as they are in the phosphonate and phosphate esters (p 102).

Approach to the phosphorus atom, which would normally occur at a tetrahedral face opposite the most electronegative or weakly bonded substituent (CH_2Ph in **XXXIV**), is postulated to occur at the tetrahedral face opposite the bulky group when extensive steric shielding is present.

For the case under consideration, a trigonal bipyramid with an axially oriented *tert*-butyl group results from **XXXIII.** The *tert*-butyl group is presumed to be effective in shielding the phosphorus atom from significant contact by the attacking hydroxide ion approaching at tetrahedral faces in which

Table 2.6.   Yields From Alkaline Hydrolysis of
Compound, $R_1R_2R_3R_4P^+$

Entry	R_1	R_2	R_3	R_4
1	tert-Bu	α–Np[a]	Ph	CH$_2$Ph
2	Me	α–Np	Ph	CH$_2$Ph
3	tert-Bu	α–Np	Ph	α–NpMe
4	Ph	α–Np	Ph	CH$_2$Ph
5	α–NpMe	α–Np	Ph	CH$_2$Ph

[a] α-Np is α-naphthyl.

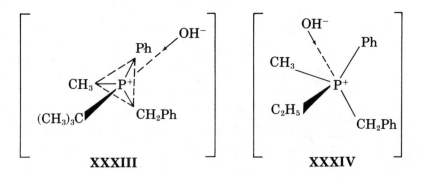

it resides. Thus, kinetic control is assumed to be operating in forming the initial trigonal bipyramidal intermediate.

Deprotonation (as suggested by McEwen and co–workers (51) for related species), followed by or accompanying pseudorotation (63), brings the benzyl group to an axial position. Loss of the latter group gives the product phosphine oxide with retention of configuration. The pathway is similar to that described in detail for **XXVI** (p 115) and shown in general form in Figure 2.1a.

Introduction of an ethoxy group, which is an excellent leaving group, in place of the benzyl group in **XXXIII**, leads (64) to nearly complete inversion of configuration following the generally "observed" route for acyclic derivatives. Thus, the use of a more electronegative group with its greater axiophilic character may override the steric control exerted by the tert-butyl group.

Corfield, De'Ath, and Trippett (65) reported yields of hydrocarbons obtained on hydrolyzing related sterically crowded phosphonium salts (Table 2.6). Interestingly, these results show that the most sterically crowded salts (Entries 1 and 5) give the least amount of toluene. This suggests that competitive loss of the bulkiest groups, α-Np and α-NpMe, respectively, presumed to be axially oriented in the formation of the initial intermediate under kinetic control, is favored over pseudorotation. The latter process would allow the better leaving group (benzyl) access to an axial site.

Sterically Crowded Phosphonium Salts (65)

Benzene	Toluene	Naphthalene	2-Methyl-Naphthalene
14	45	32.5	—
4.8	70	8.2	—
18	—	9.4	48
5	85	4.7	—
—	21	1	59

Journal of the Chemical Society

Observations similar to these were reported by Luckenbach (66, 67, 68), who demonstrated that substitution of C_2H_5 in **XXXIV** by larger ligands tended to reduce the amount of configurational inversion in the alkaline hydrolysis of the chiral phosphonium salt. Examination of the trend in Table 2.7 shows a general correlation between increasing bulkiness of Substituent R and a decline in the percentage of inversion. Accordingly, as the steric size of R is increased, competitive formation of the initial intermediate with the attacking hydroxide ion oriented opposite the larger R group is enhanced relative to entry of the hydroxide ion opposite the departing benzyl group.

The direct correlation obtained between steric crowding and retention of configuration in this nicely graded series of phosphonium salts (51, 63–68) lends support to the mechanistic interpretation presented.

As an additional factor influencing stereochemistry, Luckenbach showed that the reaction medium also has an effect. Results on the alkaline cleavage of phosphonium salts of the type (68) showed differences in stereoselectivity

$$\left[\begin{array}{c} CH_3 \\ | \\ C_6H_5-P-R \\ | \\ tert\text{-}C_4H_9 \end{array} \right]^+$$

in heterogeneous aqueous solution compared with a homogeneous medium of aqueous ethanol. The stereochemical changes appear to correlate with changes in the efficiency of the leaving group brought on by different degrees of solvation in the two media.

With reference to reaction kinetics for structures such as **XXXIII**, Trippett and co-workers (47, 63) observed little effect on the rate of alkaline hydrolysis when one *tert*-butyl group was present. However, two *tert*-butyl groups produced a large rate retardation. Alkaline hydrolysis of **XXXV** proceeds about one-fifth as fast (63) as **XXXVI**, while the di-*tert*-butylphosphonium salt **XXXVII** is extremely resistant. After 11 days at 100°C in 1N

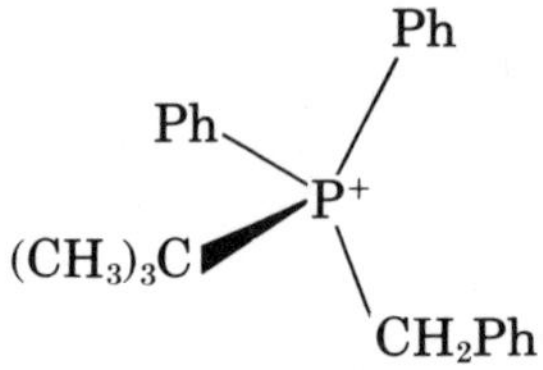

XXXV 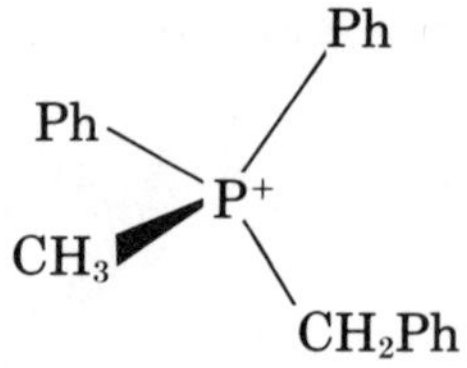   **XXXVI** 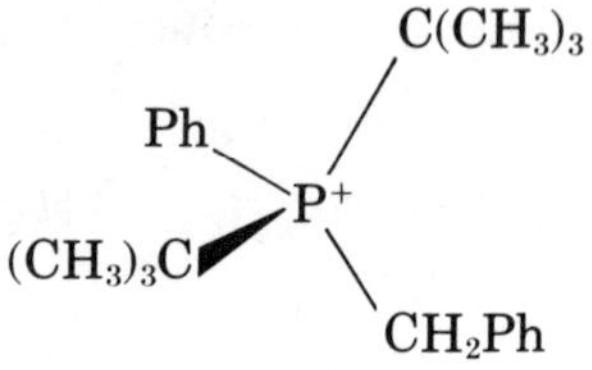   **XXXVII**

sodium hydroxide (90% ethanolic solution), 21% of this salt (**XXXVII**) remained unchanged. Likewise, for acyclic phosphinate esters the rate constant (47) for alkaline hydrolysis of $(CH_3)_3C \overset{O}{\underset{\|}{P}}(OC_2H_5)_2$ is almost a factor of 10^3 greater at 120°C than that for $[(CH_3)_3C]_2\overset{O}{\underset{\|}{P}}OC_2H_5$ (Table 2.4).

Similar to the shielding argument presented for **XXXIII**, attack by the hydroxide ion trans to the *tert*-butyl group would lead to little steric hindrance. However, with two *tert*-butyl groups present, hindrance to the approach of the attacking hydroxide ion in the transition state **XXXVIIIa** leading to

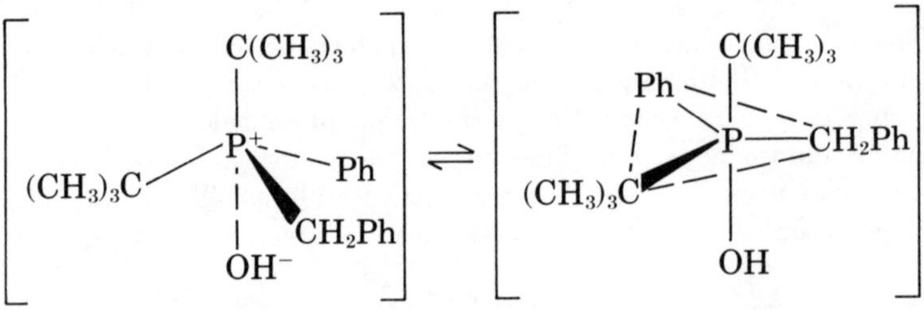

XXXVIIIa   **XXXVIIIb**

Intermediate **XXXVIIIb** should result in a reduced reaction rate. Hydroxide ion attack in this manner would be in keeping with locating the entering group in an axial position. It is interesting that the major phosphorus-containing product here is benzyl-*tert*-butylphenylphosphine and isobutene, implying a possible steric inhibition of pseudorotation of **XXXVIIIb** if it is correctly assumed that **XXXVIIIb** forms in the first place. The reaction products are consistent with a Hofmann elimination (69).

Electronic Effects. In a study of the alkaline decomposition of a series of meta- and para-substituted tetrabenzylphosphonium halides, $(y\text{-}PhCH_2)$ $(PhCH_2)_3P^+X^-$, McEwen et al. (52) reported that all members reacted much faster than **XXXIV**. To rationalize the rate retardation in the di-*tert*-butylphosphonium salt **XXXVII** and the rate acceleration reported for the tetrabenzylphosphonium halides, approach of the attacking hydroxide ion must be more favorable in the latter system. Two differences are noted that might

Table 2.7. Stereochemistry in the Cleavage of Chiral Phosphonium Salts by Aqueous Alkali (66)

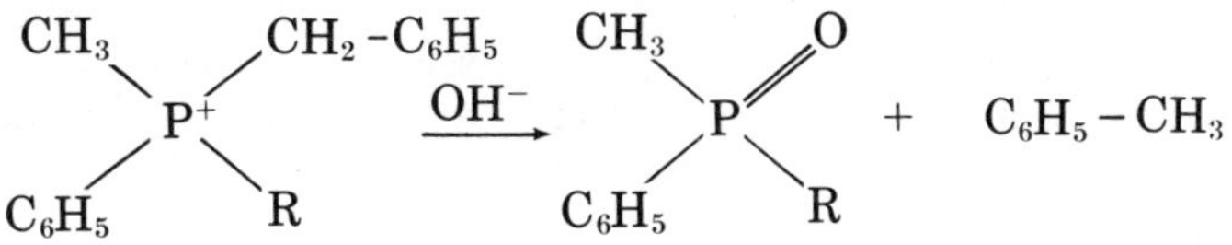

$$
\begin{array}{c}
CH_3 \diagdown \quad \diagup CH_2-C_6H_5 \\
P^+ \\
C_6H_5 \diagup \quad \diagdown R
\end{array}
\xrightarrow{OH^-}
\begin{array}{c}
CH_3 \diagdown \quad \diagup O \\
P \\
C_6H_5 \diagup \quad \diagdown R
\end{array}
+ \quad C_6H_5-CH_3
$$

R	Inversion of Configuration[a] (%)
iso-C₃H₇	97
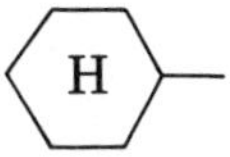 (H)—	100
(CH₃)₃C-CH₂⁻	100
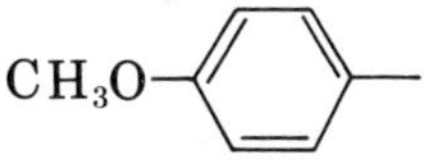 CH₃O—⟨◯⟩—	86
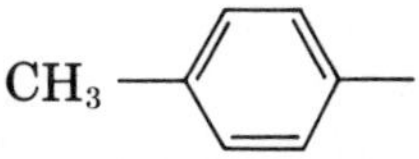 CH₃—⟨◯⟩—	58
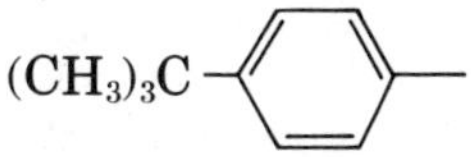 (CH₃)₃C—⟨◯⟩—	47
α-Np	28

[a] The percent racemization is defined as rac = 100 − inv, where inv is the percent inversion based on the ratio of the optical purity of the phosphine oxide and the phosphonium salt. Therefore, the percentage of molecules inverted is inv + 0.5 rac, and that retained is 0.5 rac.

Phosphorus

bring this about. First, the benzyl group probably offers less steric hindrance than a *tert*-butyl group to the approach of the attacking hydroxide ion, particularly since rotations about the P–C bond (as shown) may act to reduce

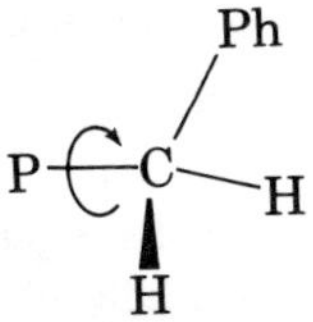

$$
\begin{array}{c}
\qquad\qquad Ph \\
\qquad\qquad \diagup \\
P \big(C \diagdown H \\
\qquad | \\
\qquad H
\end{array}
$$

the steric interaction. Second, the preponderance of more electronegative groups in the tetrabenzyl system will increase the nucleophilicity of the phosphorus atom and thus render it more susceptible to attack by hydroxide ion (52).

It was observed that rate accelerations of the order of 10 to 10^2 were obtained with electron-withdrawing substituents (y) attached to meta and para positions of a single benzyl group when this group, so substituted, was the departing group. If the leaving group was an unsubstituted benzyl group and the nondeparting group the substituted benzyl group, the rate enhancement was much more modest, of the order of 1–5.

Additional evidence for the effect of electron-withdrawing substituents on the rate of reaction is found in the much greater rate of alkaline hydrolysis reported (70) for tri(2-furyl)methylphosphonium iodide (**XXXIX**, R = CH_3, X = I) and benzyltri(2-furyl)phosphonium bromide (**XXXIX**, R = CH_2Ph, X = Br), giving di(2-furyl)methylphosphine oxide (**XL**, R = CH_3) and benzyldi(2-furyl)phosphine oxide (**XL**, R = CH_2Ph), respectively, compared with the analogous reaction of methyltriphenylphosphonium iodide (giving methyldiphenylphosphine oxide and benzene). The rate of hydrolysis

XXXIX　　　　　　**XL**

of the trifuryl derivatives is about 10^{10}–10^{11} times faster than the phenyl analog. Furthermore, hydrolytic cleavage of **XXXIX** proceeds about 10^2–10^3 times faster than the hydrolysis of the corresponding tri(2-thienyl)phosphonium salts (**XLI**, R = CH_3, X = I and R = CH_2Ph, X = Br), giving the oxides (**XLII**) and thiophene.

Allen et al. (70) suggested that the greater electron-withdrawing character of the heteroaryl substituents, which increases from 2-thienyl to 2-furyl, is responsible for vastly increased hydrolytic rates of **XXXIX** and **XLI** com-

XLI　　　　　　**XLII**

pared with the hydrolysis of methyltriphenylphosphonium iodide by providing greater concentrations of pentacoordinate intermediates, R_4POH and R_4PO^-. The latter are formed in rapid preequilibria steps (52) (*see* mechanism outlined on p 115 for **XXVI**). Also, the heteroaryl anions eliminated in the rate-determining step are considered to be more stable than the phenyl anion. As a consequence, the overall reaction rates appear to be a result of a multiplicative effect.

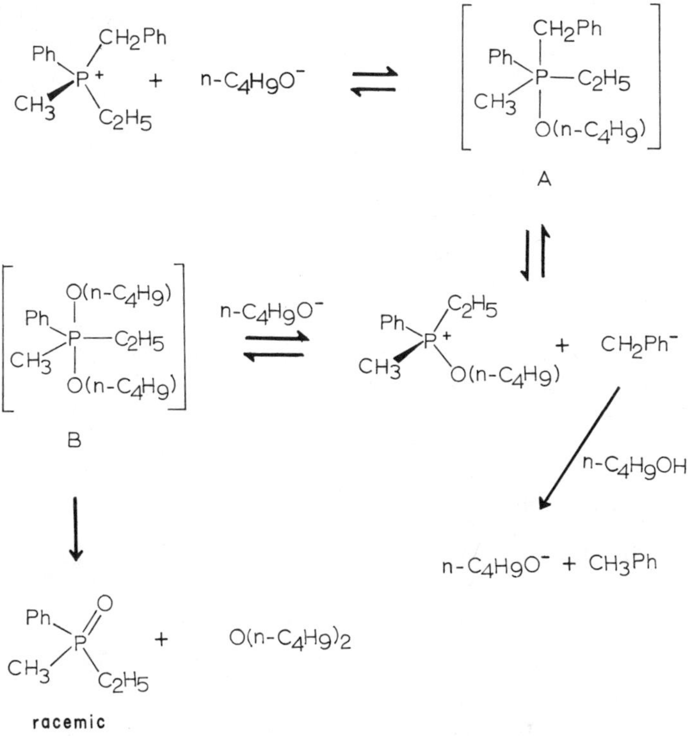

Journal of the American Chemical Society

Figure 2.5. Proposed mechanism (71) for the formation of racemic phosphine oxide from the decomposition of chiral methylethylphenylbenzylphosphonium iodide by n-butoxide

The importance of preequilibria in these hydrolyses, which are thought to follow the mechanism outlined by McEwen et al. (52), was demonstrated by observing (70) a marked shift in the ^{31}P NMR resonance on addition of sodium methoxide to the furyl derivative **XXXIX** (R = CH$_3$) in absolute methanol ($\delta = +91.7$ ppm, relative to 85% H$_3$PO$_4$). The shift is in the region characteristic of five-coordinate phosphorus (*see* Volume I, Tables 3.4 and 3.12). No significant positive shift in the ^{31}P signal occurred with either the thienyl derivative, **XLI** (R = CH$_3$), or the methyltriphenyl derivative.

An additional factor that may contribute appreciably to rate effects and for which adequate assessment has not been made involves pi bonding. Studies show that pi bonding concentrates in equatorial bonds attached to phosphorus in trigonal bipyramids (p 29). This effect should be particularly important in the furyl and thienyl derivatives. An x-ray study of a related pyrrole derivative (Volume I, p 25) revealed an unusually short equatorial P–C bond, indicative of electron delocalization from the ring into the P–C linkage.

Table 2.8.   Stereochemistry of Nucleophilic Salts and Their

Entry	Compound	Nucleophile/ Leaving group
1	Ph, *tert*-Bu—P^+—SMe^b, OMe	OH$^-$/SMe$^-$
		OH$^-$/OMe$^-$
2	Ph, Me—P^+—SR$_2$^{c}, OR$_1$	OH$^-$/SR$_2$$^-$ and OH$^-$/OR$_1$$^-$

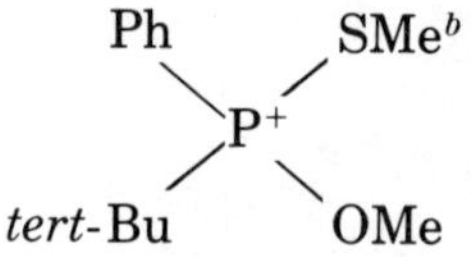

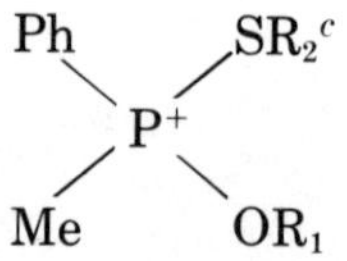

	R_1	R_2
a	Me	Me
b	Me	Et
c	Me	iso-Pr
d	Et	Me
e	Et	Et
f	Et	iso-Pr
g	iso-Pr	Me
h	iso-Pr	Et
i	Men	Me^e

| 3 | Ph, MeS—P(=O)—OMen | OMe$^-$/SMe$^-$ |

| 4 | Ph, MeS—P(=O)—OR | MeMgI/SMe$^-$ and MeMgI/OR$^-$ |

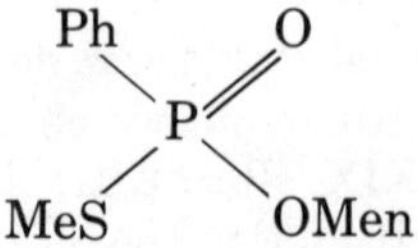

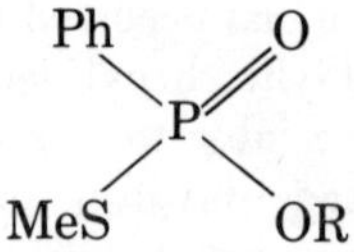

	R
a	Me
b	iso-Pr
c	Men

Displacement on Thio–Containing Acyclic Phosphonium Neutral Esters

Product (%)	*Ret-Inv*[a]	*Ref*

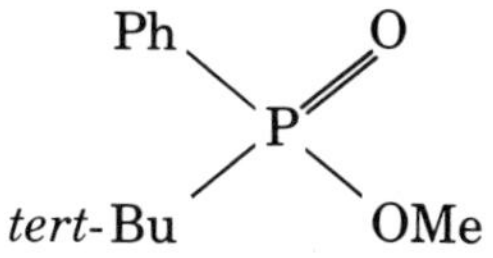

Ret — 72, 73

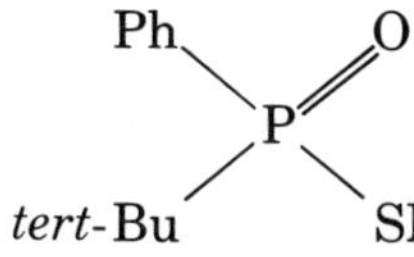

22%   —

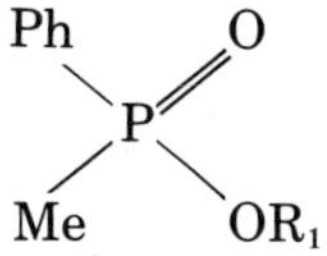

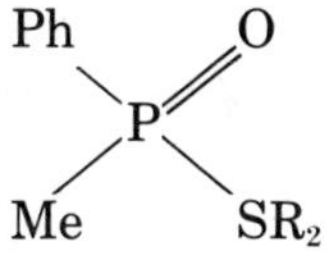

73

$\%^d$	$\%^d$
65 Ret	35 Inv
67	33
67	33
79	21
81	19
80	20
91	9
90	10
>95 Ret	<5

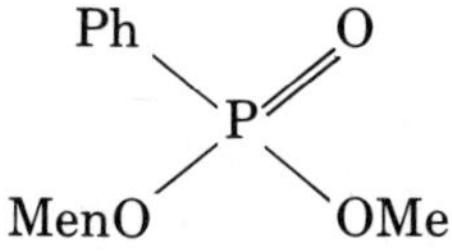

Inv   74

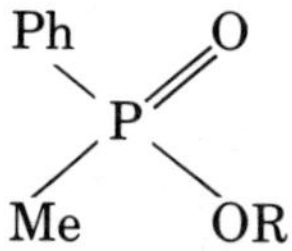

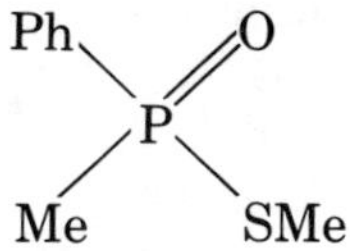

75

$\%$	$\%$
25 Ret	75 Inv
80 Ret	20 Inv
100 Ret	0

Table 2.8.

Entry	Compound	Nucleophile/ Leaving group

5	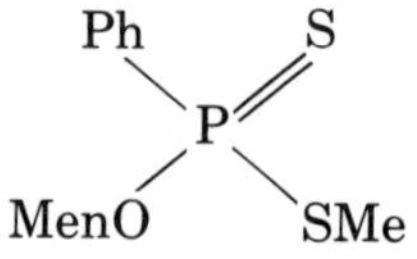 Ph, S, P, MenO, SMe	OH^-/SMe^-
6	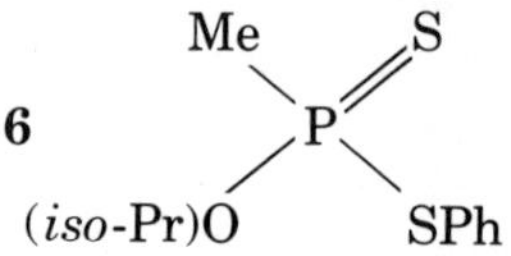 Me, S, P, $(iso\text{-Pr})O$, SPh	OH^-/SPh^-
7	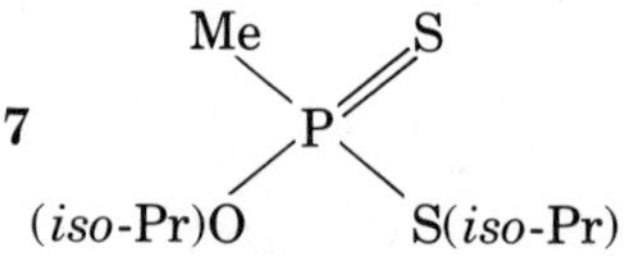 Me, S, P, $(iso\text{-Pr})O$, S$(iso\text{-Pr})$	$OH^-/S(iso\text{-Pr})^-$
8	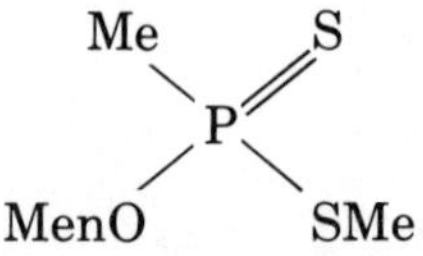 Me, S, P, MenO, SMe	OH^-/SMe^-

[a] Predominant retention or inversion of the chiral center is indicated.
[b] In addition to cleavage of the P–S bond for Entry 1, De Bruin and Johnson (73) reported 22% cleavage of the methoxy group to form methyl tert-butylphenylphosphinothiolate but, unfortunately, did not investigate the resultant stereochemistry. Hence, the controlling influence in determining which group occupies the axial position in the initial intermediate is uncertain.
[c] The stereochemistry of the products were investigated only for the entries indicated (**2a** and **2i**).

In terms of electron-pair repulsion theory (p 4), equatorial concentration of pi bonding in the trigonal biypramidal intermediate would shift electron density toward the phosphorus atom and be contributory in expelling the axial anionic ligand. Although some effort (65) has been made to correlate the effects of pi bonding on reaction mechanism, the limited amount of data available has not led to a satisfactory assessment (see p 148).

An interesting result was obtained by McEwen and co-workers (71) who observed that 1-butoxide in butanol caused cleavage of chiral methylethylphenylbenzylphosphonium iodide to give methylethylphenylphosphine oxide with more than 90% racemization. This contrasts with the "normal" inversion route found for the alkaline hydrolysis of this compound (p 119). The result does not seem to be associated with a steric effect, as was the case when tert-butyl groups were present (p 118). The mechanism postulated

Continued

Product (%)	Ret-Inv[a]	Ref
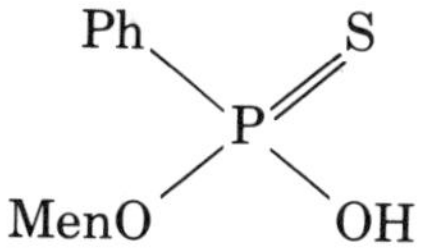 Ph, S, P, MenO, OH	Inv	76
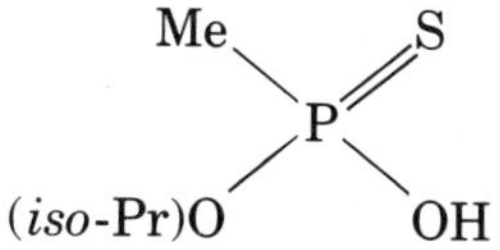 Me, S, P, (*iso*-Pr)O, OH	Ret	77
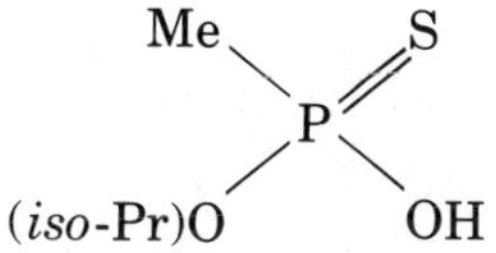 Me, S, P, (*iso*-Pr)O, OH	Ret	77
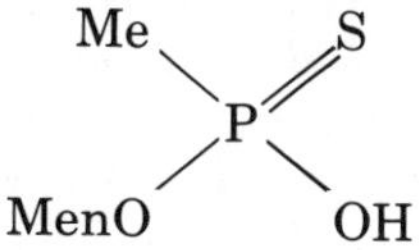 Me, S, P, MenO, OH	Inv	78

[a] The percentage of products obtained for Entries **2**, which are sensitive only to the alkyl substituent R_1, appears to be a function of the leaving ability of the alkoxy group: OMe > OEt > O-iso-Pr (*73*). With the poorer leaving group, there may be more time for the initial intermediate having the alkoxy ligand in an axial site to pseudorotate and allow competitive loss of the alkylthio group from the axial position in the newly formed intermediate.

[e] Predominant retention for Entry **2i** also was reported by De'Ath et al. (*72*).

(Figure 2.5) involves an achiral trigonal bipyramidal intermediate. Product formation from the latter accounts for the extensive racemization observed. Substantiation for the proposed mechanism was obtained from the observance of the rapid formation of toluene. Compared with the 8% yield of di-*n*-butyl ether, toluene was formed in near quantitative amounts (*71*).

Nucleophilic Displacements of Acyclic Derivatives Containing Thio Groups

In the case of hydrolysis of the cyclic thioate, **XXII** (R = OCH_3, p 110), the rate factor was greatly enhanced relative to its oxy analog, and P–S ring cleavage dominated over most of the pH range, i.e., above pH of 1.5. The greater leaving ability of the alkylthio ligand (*72, 73*) and the weakness of

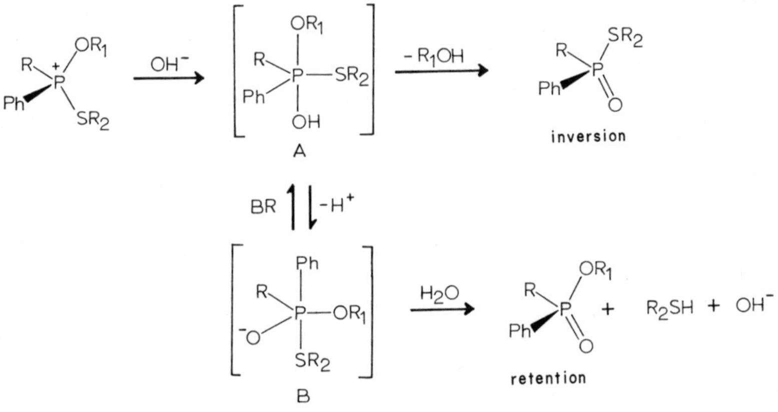

Figure 2.6. Competitive loss of alkoxy and alkylthio ligands in the alkaline hydrolysis of alkoxy (alkylthio)phosphonium salts. The scheme is formally analogous to that in Figure 2.1.

the P–S bond compared with these properties for the oxy analogs were cited as the primary factors. The interplay among the various determinants entering into mechanistic consideration for reactions of thio derivatives has been studied in the absence of ring constraints. These studies center largely on reactions involving nucleophilic attack on acyclic phosphonium salts and their neutral esters. A summary of some of the results (72–78) of the stereochemical outcome for reactions using chiral centers is given in Table 2.8.

Normally, as depicted for the acyclic phosphonium salt, **XXXIV** (p 119), the attacking nucleophile is expected to enter a tetrahedral face opposite the most electronegative substituent. This process is postulated (72, 73) for Entries **1** and **2** of Table 2.8, leading to Intermediate **A** of Figure 2.6. It is noted that similar to Entry **1**, the presence of a *tert*-butyl group in the phosphonium salt, **XXXIII** (p 119), was postulated to cause steric shielding and give rise to kinetic control—i.e., orientation of the attacking group at a tetrahedral face opposite the bulky group. However, substitution of the more electronegative ethoxy group in place of benzyl, which represents the leaving group in the hydrolysis of the latter salt, gives results (p 120) that are consistent again with OH⁻ attack opposite the most electronegative substituent. Thus, the formation of Intermediate **A** (Figure 2.6) for Entry **1** appears reasonable although not proven (*see* Footnote *b* to Table 2.8).

Subseqeunt loss of the axial alkyl group from Intermediate **A** will yield inversion of configuration. Alternatively, pseudorotation of this intermediate may take place and, in the process, bring the thio ligand to a departing axial position (Intermediate **B** of Figure 2.6). Loss of this thio ligand by this route, as postulated for Entry **1** (72, 73), results in retention of configuration. Even though the alkoxy group is more electronegative than the alkylthio group, the latter (as previously noted) is the better leaving group. This then becomes an important factor when discussing mechanistic routes for thio derivatives. The decrease in the amount of inversion reported (73) for Entries 2a–2h as the alkoxy group is changed from OMe to OEt to O-iso-Pr parallels the decrease in leaving ability of these groups, in that order, as measured by the acidity of the corresponding alcohol in aqueous medium (79). This observation tends to substantiate the mechanism in Figure 2.6 where competitive loss of the alkylthio group from **B** is expected to increase as the alkoxy group becomes less able to depart from **A**. A similar trend is present regarding product stereochemistry for the Entries **4a**, **4b**, and **4c**.

However, the above reasoning fails to account for the inversion of configuration that takes place (74) in the methanolysis of Entry **3**, which closely resembles Entry **4**. Likewise, examination of the stereochemical consequences of reactions involving phosphonodithioates, Entries **5**, **6**, **7**, and **8** (76, 77, 78), reveals little rationale for the outcome either in terms of the suggested leaving ability of the thio group, iso-PrS > SMe > SPh (78), or relative apicophilicity of the alkoxy group. The unexpected result is the inversion of configuration accompanying the loss of the methylthio group for Entries **3**, **5**, and **8**. In all of these reactions, the O-menthyl group is present.

o-Menthyl

We speculate here that a steric factor is implicated. Taking advantage of Preference Rule 4 (p 88), which places the most bulky group equatorially in a trigonal bipyramid, it seems reasonable that the intermediate takes the form shown, whether the displacement is concerted or has a finite lifetime.

XLIII

Under this proposal, we presume that steric interactions render this intermediate energetically more stable than one with the O-menthyl group located in an axial position. Thus, thermodynamic control is operating. The presence of the large sulfide ion situated equatorially in the intermediate formation (**XLIII**) also favors equatorial positioning of the O-menthyl group. Carrying this argument further, increasing the steric bulk of the attacking nucleophile should favor axial orientation of the O-menthyl group and return to retention of configuration, as illustrated in Figure 2.6. This latter point, however, does not seem to have been subjected to adequate testing.

The above mechanism postulated for Entries **3**, **5**, and **8** differs from the mechanism advanced for the hydrolysis of **XXXIII** (p 119), which is thought to be under kinetic control. In the latter phosphonium salt, the substituents are all relatively close in electronegativity so that hydroxide ion attack will readily take place opposite the most sterically shielded portion of the molecule.

We further suspect, although cannot substantiate to any significant degree, that the reason Entry **2i**, (which also involves an O-menthyl group) loses the methylthio group with retention of configuration is because we are dealing with a phosphonium salt rather than a neutral ester. In the salt, electronic factors appear to have added importance. As a result, the O-menthyl group is oriented axially in the initial intermediate (**A** of Figure 2.6). Further, the attacking nucleophile for Entry **2i** relative to that for Entry **3** is more electronegative $OH^- > OMe^-$, and this feature may contribute to the operation of electronic control in intermediate formation.

A delicate balance exists among the various factors influencing the stereochemical outcome for reactions of mixed alkoxy–alkylthio phosphonium salts and their neutral esters. The previous discussion is meant to be a working hypothesis as an aid in guiding future work for reactions of this very interesting class of substances.

Substituent Effects in Cyclic Systems

Four-Membered Rings. NEOPENTYL EFFECT. Hawes and Trippett (*47*) reported that the alkaline hydrolysis of the cyclic phosphinate ester, **XXIII** (p 113), proceeds extremely rapidly. Bergesen (*80*) found that the alkaline hydrolysis of the related, but more methylated derivative, **XLIV**, is approximately equal to that found for the acyclic ester, $(C_2H_5)_2PO(OC_2H_5)$ (Table 2.4). Hawes and Trippett (*47*) also studied the alkaline hydrolysis of **XLIV** and obtained similar rate results (Table 2.4).

XLIV

Since the rapid hydrolysis of **XXIII** was attributed to relief of ring strain on formation of the Intermediate **XXIV**, a similar acceleration is expected in **XLIV** compared with an acyclic derivative. To account for the rate reduction, a "neopentyl" effect was suggested (*47*), involving steric hindrance between the α-methyl groups located on the ring and the attacking hydroxide ion. The argument for the rate retardation parallels that discussed on p 122 for the alkaline hydrolysis of benzylphenyl di-*tert*-butylphosphonium bromide, **XXXVII**.

In the corresponding phosphinate ester, 1-methoxy-2,2,3,4,4-pentamethylphosphetane 1-oxide (**XLV**), Cremer and Trivedi (*81*) showed that attack of methoxide ion on either the cis or trans isomers, containing deuterium-labeled methoxy groups, takes place with retention of configuration and without isomer crossover (Figure 2.7). Hence, the normal retention pathway is followed (Figure 2.1).

A number of other substitution reactions at phosphorus in 2,2,3,4,4-pentamethylphosphetanes have been shown to follow a retention scheme. Some of these (*82*) are summarized in the following cycle (*see* also Figure 2.14).

Journal of the Chemical Society (*82*)

The fact that both the structure of the phenylphosphine oxide (*83*) isomer (**XLVI**) and the acid chloride (unpublished x-ray work by M. Haque, Ref *85*) isomer used in these studies have been established as the trans configuration by single crystal x-ray analysis suggests a retention mechanism for the entire cycle. (As Cremer and Trevedi point out (*84*), the word "cis" in the paper by Ul-Haque and Caughlan (*83*) should be "trans." This is apparent as well from the figure in the latter paper.) *See* p 113 for a discussion of estimated relative energies of intermediates for reactions involving four-membered rings. The yield of the phosphine oxide, **XLVI**, in the cycle is low because of a competing reaction of phenyllithium with this oxide (*82*).

RING EXPANSION AND RING OPENING. Treatment of Phenylphosphetane Oxide **XLVI** separately with phenyllithium in ether (*82*) followed by protonation of the product with water gives the secondary phosphine oxide, **XLVIII**, (Figure 2.8). Deuterium labeling of the two phenyl nuclei shows

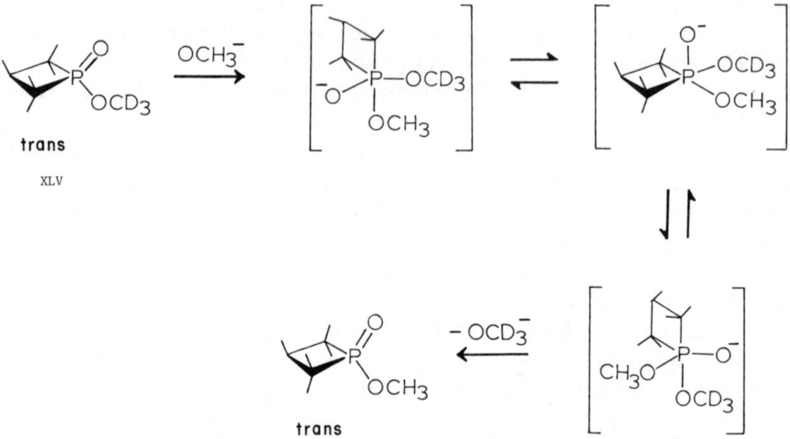

Figure 2.7. Retention pathway for trans-1-methoxy-2,2,3,4,4-pentamethyl-*phosphetane 1-oxide,* **XLV** *(81). Although the trans isomer is illustrated, the cis isomer follows the same route. A line is used to denote the presence of a methyl group. A square pyramidal transition state is included in this variant of the proposed scheme (81) (see p 87).*

that they become equivalent (*82*) during the reaction. The latter suggests (*82*) pseudorotation in a ring expansion mechanism, giving Intermediate **XLVII**. (Hawes and Trippett (*82*) postulated that the pseudorotated conformer has the electropositive group ($-O^-$) axially oriented. However, the location of the more electronegative phenyl group in an axial position is preferred in **XLVII**. Equivalence of the phenyl groups is obtained by way of **B** under the mechanistic variation presented here.) Possibly the ring expansion is brought about by steric crowding due to the attacking phenide ion, causing enhancement of the neopentyl effect relative to that present in **XLIV**. Entrance of the phenyl ligand at an axial site should produce considerable steric hindrance in the initial intermediate, **A**. Pseudorotation through the transition state (**B**) to give **XLVII** allows the phenyl groups to achieve equivalence. Axial $C(CH_3)_2$ migration follows from **XLVII** to give **C** with ring retention. However, the ring is not retained in the final product (**D**) since it cannot be protonated in this reaction mixture (*82*). Aqueous treatment leads to the secondary phosphine oxide, **XLVIII**.

A related situation exists in the alkaline hydrolysis of the phosphetanium salt, **XLIX**, which undergoes ring expansion (*56, 86*) and gives Phosphine Oxide **LI**. Following a mechanistic approach (Figure 2.9) similar to that for **XLVI,** the phenyl group again enters into the steric crowding enhancing the neopentyl effect. The latter may be causative in the rupture of the P–phenyl bond at the expense of or during the pseudorotational process. The latter process is not shown in Figure 2.9 but would follow from Intermediate **L**. Thus, migration of the axial $C(CH_3)_2$ of the ring is postulated (*86*), giving

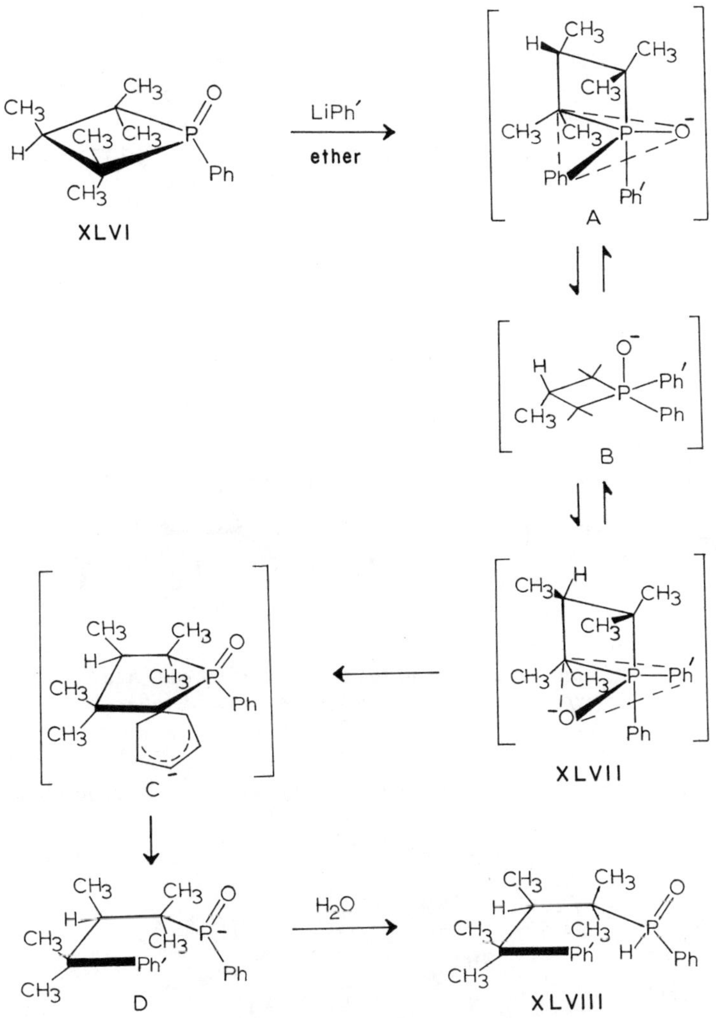

*Figure 2.8. Formation of the secondary phosphine oxide, **XLVIII**, with retention of configuration by a ring expansion mechanism resulting from the treatment of 2,2,3,4,4-pentamethyl-1-phenylphosphetane oxide **XLVI**, with phenyllithium*

the cyclohexadienyl anion. On protonation the nonconjugated isomer, **LI**, is formed, having a spirobicyclic structure.

Alkaline hydrolysis of the corresponding benzyl derivative (*56, 82*), **LII**, proceeds by the more "normal" route to give the phosphine oxide, **XLVI**, as the major product, the latter with apparent retention of configuration. The reaction is shown giving the major isomer, mp 127°C, which is the one that has been established by x-ray analysis (*83*) to have the methyl and phenyl

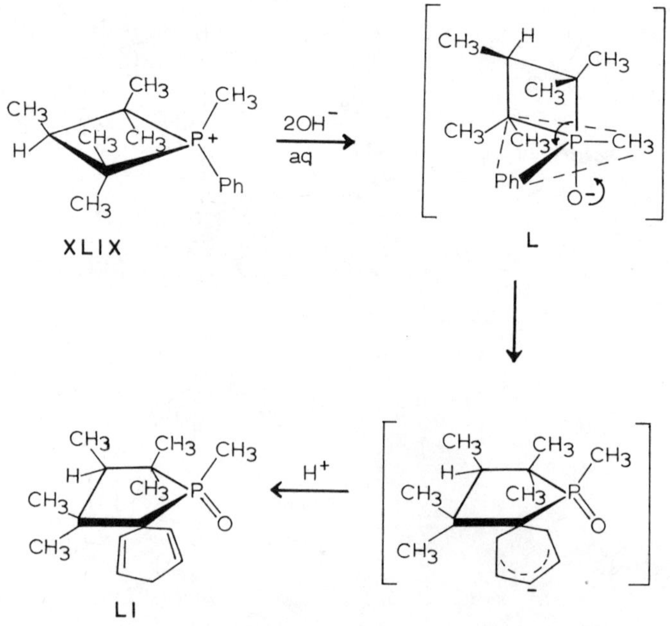

Figure 2.9.   Alkaline hydrolysis of the phosphetanium salt, **XLIX**, *with ring expansion to give the phosphine oxide,* **LI** *(86)*

groups on opposite sides of the ring (p 133). The retention mechanism is

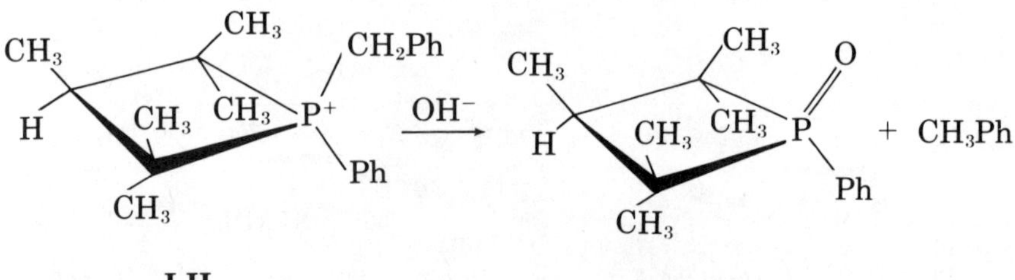

LII   XLVI

postulated, leading to loss of the benzyl ion from an axial position after pseudorotation of the pentacoordinate intermediate initially formed. Hawes and Trippett (*82*) assumed pseudorotation to be less probable than equatorial loss of the benzyl anion since they considered the O⁻ ligand to be the most electronegative group present. In the pseudorotated structure, this group would be moved to an equatorial position. However, as Luckenbach has pointed out (*87*), the O⁻ ligand is a very strong electron-donating group. *See* Ref *88.*

The steric hindrance discussed for **L** also would be present in the latter intermediate for the benzyl derivative. However, the presence of the benzyl group apparently exerts a controlling influence on the course of the reaction largely due to the greater stability of its anion (*56, 82*) relative to that of the phenyl anion. Consequently, the benzyl group would depart more readily from the pseudorotated conformer, **LIV**, formed from the initial intermediate, **LIII**. As a result, ring expansion and ring opening presumably are less favorable processes in this case.

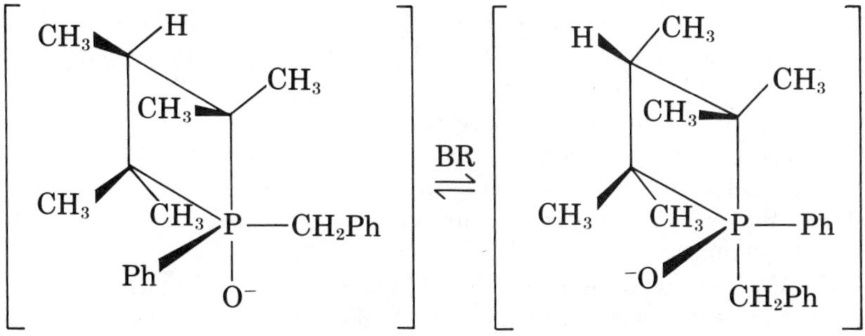

LIII   **LIV**

If the phenyl group in **LII** is replaced by a *tert*-butyl group, some ring cleavage is noted (*56*) during alkaline decomposition, giving the phosphine oxide, although the cyclic oxide, 1-*tert*-butyl-2,2,3,4,4-pentamethylphosphetane 1-oxide, is still the major product. The formation of some acyclic product is suggestive of increased steric effects between the bulky *tert*-butyl group and the adjacent, *gem*-dimethyl substituent located at an axial position of the initial trigonal-bipyramidal intermediate. Both weakening of the axial P–C ring bond and inhibition of the pseudorotational process are expected to contribute to ring cleavage (cf. p 122).

Interestingly, both Cremer et al. (*89*) and Trippett and co-workers (*90*), reported that partial inversion takes place in the alkaline hydrolysis of the cis isomer of the phosphetanium salt, **LII**. No matter what isomer composition is taken for the initial mixture, the product composition is predominantly the trans isomer (*89*). Partial inversion also is reported in the alkaline hydrolysis of Phosphetanium Salts **LV** and **LVI** (*56, 90*) and in

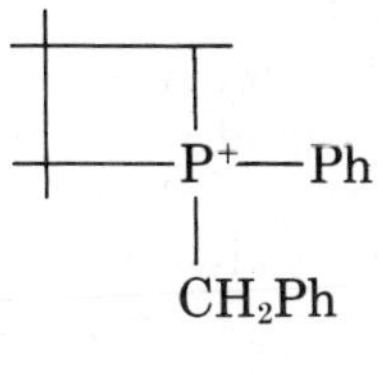

(a) R = R' = CH$_2$Ph (*56*)

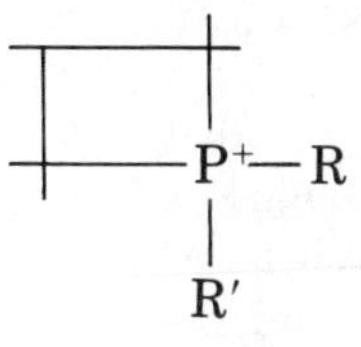

(b) R = CH$_3$, R' = CH$_2$Ph (*56*)

XLIX R = Ph, R' = CH$_3$ (*56, 90*)

LV   **LVI**

olefin synthesis (*90*) of the cis isomer of **LII**. However, the optically active structure:

$$\overset{\displaystyle \underset{|}{}}{\underset{\underset{\displaystyle CH_2Ph}{|}}{\text{---}P^+\text{---}Ph}}$$

shows no evidence for partial inversion (90). The fact that the pure cis- and trans-phosphetane oxides formed from **LII** and **LVI(b)** do not interconvert under the conditions used in alkaline hydrolysis implies that isomer cross-over occurs at an intermediate stage in the reaction sequence. Cremer et al. (89) favor a pseudorotation mechanism that involves at least three steps before decomposition to product (Figure 2.10). Trippett and co-workers (90) postulate that isomer crossover occurs via ylides in both alkaline hydrolysis and olefin synthesis.

On the one hand, it seems less likely that successive pseudorotations occur beyond the formation of **c** to give higher energy intermediates such as **e** and **f**; the pseudorotational barrier is estimated as $\sim$7 kcal/mol when R = Ph and $\sim$8 kcal/mol when R = Me. On the other hand, it has been established that the Wittig reaction takes place with retention of configureation (53), at least for acyclic phosphonium salts. However, the phosphonium salts under discussion contain highly strained rings with substituents capable of exerting large steric effects. We have seen that the four-membered ring is prone to open in alkaline hydrolysis when the phosphonium salt contains poor departing groups—e.g., **XLIX**. It is suspected that even with the benzyl group present, the cis isomer exerts enough additional steric hindrance over that encountered in the trans configuration during hydroxide decomposition that P–C ring bond rupture occurs, leading to isomer equilibration as suggested by Trippett and co-workers (90).

Once isomer equilibration is achieved, alkaline hydrolysis and Wittig olefin synthesis can proceed according to the retention pathway (cf. Figure 2.1a for the general scheme or more specifically Figure 2.7 for details of the retention process for the analogous methoxide exchange). The olefin synthesis with **LII** is illustrated in Figure 2.11. The latter scheme is analogous to that postulated (51, 91) for the Wittig reaction with acyclic chiral phosphonium salts proceeding with retention of configuration (51, 53). It is interesting that the crystal structure (92) of a Wittig intermediate is approximately trigonal bipyramidal with an unusually long P–O axial bond (Volume I, p 25). The latter bond distance is 2.01 Å, which compares with about 1.8 Å for the longest distance obtained for phosphorane derivatives (6, 7). These results imply that the intermediate in a Wittig synthesis is not a true betaine (92) but somewhere between the latter and the normally observed trigonal bipyramidal phosphorane structure (6, 7).

Substitution of ethoxy (a better leaving group), for the benzyl group in the phosphetanium salt, **LII**, results in complete retention of configuration independent of which isomer is involved in alkaline hydrolysis. Isomer crossover is predictably less likely in this derivative (**LVII**), either via multiple

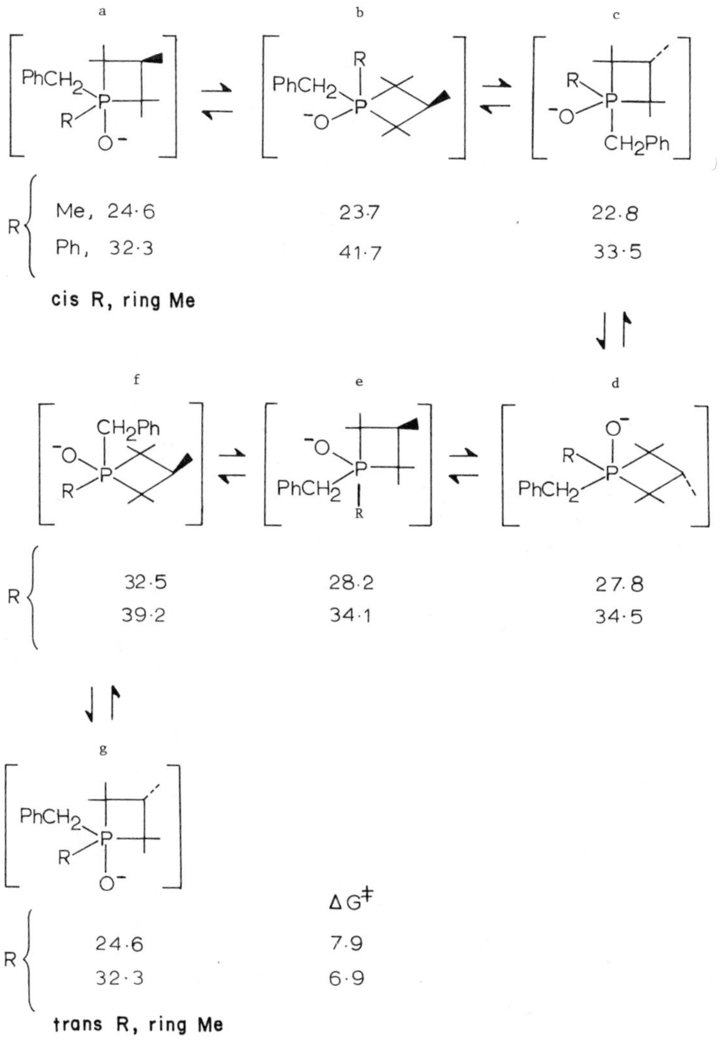

Figure 2.10. A specific mechanism for isomer equilibration is given here based on the proposed operation of three successive pseudorotations in the alkaline decomposition of sterically hindered cis phosphetanium salts (89). Numbers below structures are estimates of relative isomer energies; however, no estimate of the steric effect of either the 3-methyl or the benzyl group is included (see p 38). If a reasonable steric factor is included (see p 37), approximately three each for the interactions $(CH_3)_2C-P-CH_2Ph$ and $Ph-P-CH_2Ph$, the estimated pseudorotational barrier increases to ~6 kcal/mol when R = Me and ~9 kcal/mol when R = Ph.

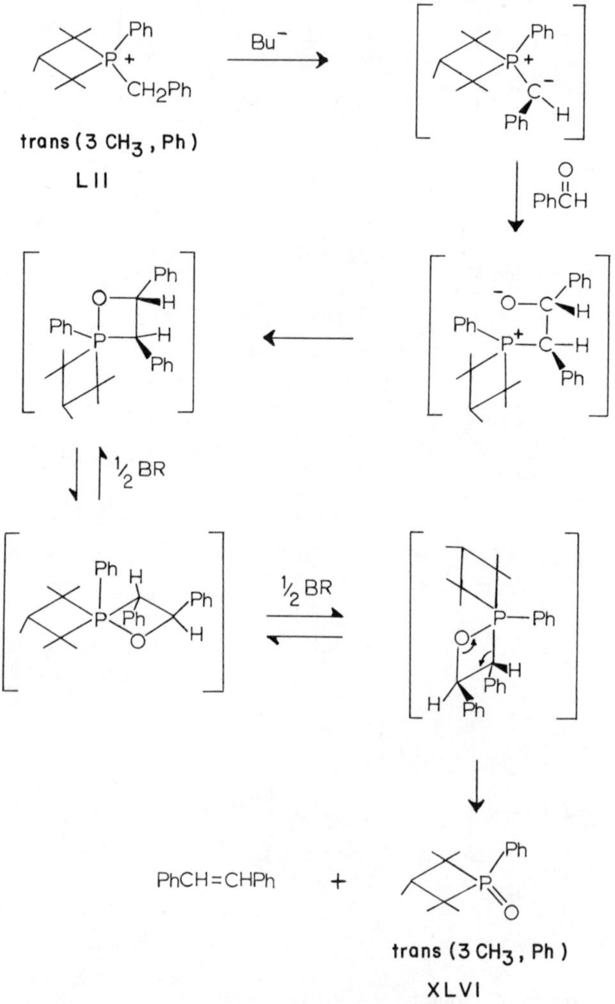

Figure 2.11. Retention pathway for the Wittig olefin synthesis illustrated using the trans form of the phosphetanium salt, **LII.** *See text for a discussion of isomer equilibration.*

pseudorotations (Figure 2.10, R = Ph) or P–C ring bond cleavage in ylide formation since rapid departure of the ethoxy group from the pseudorotated Intermediate **c** of Figure 2.10 should render these other processes less competitive. A number of other reactions have been proposed to follow mechanisms similar to that for the Wittig reaction. These should be consulted by the interested reader (*93–98*).

LVII

In the alkaline decomposition of the less sterically shielded Phosphe-tanium Salt **LVIII**, analogous to **XLIX** (p 136) but ontaining two less α-methyl ring substituents, predictable differences are found (*56*). The rate of hydrolysis of the former is greater than that of the latter, and the reaction of **LVIII** proceeds with ring opening, leading to Oxide **LIX** rather than ring expansion, as observed for **XLIX**. The reduced steric interaction undoubtedly

LVIII   **LIX**

favors a more facile pseudorotational process as well as some degree of stabilization of the P–Ph bond relative to that in the initial intermediate formed from **XLIX**.

A parallel in mechanistic behavior seems to exist to some degree between the alkaline hydrolysis of acyclic phosphonium salts containing one and two *tert*-butyl groups (p 119ff) and the hydrolysis for the respective cyclic systems here, **LVIII** and **XLIX**. Similar to the suggestion that steric shielding provided by a *tert*-butyl group orients attack by hydroxide ion on **XXXIII** at an opposite tetrahedral face, the initial intermediate proposed by Cremer et al. (*56*) contains the ring —C(CH₃)₂ group at an axial site. (Location of the *gem*-dimethyl groups, as formulated by Hawes and Trippett (*47, 48*), at an axial site in the intermediate is consistent with the operation of kinetic control that appears to take place when a bulky group is present. This effect is discussed in more detail on p 133 for a related phosphinate ester, **XXXIII**, shown on p 119. Pseudorotation followed by ring opening results in a retention pathway (Figure 2.12). However, the stereochemical course of this reaction does not seem to have been studied. A related rate comparison between phosphinate esters, structured similar to **LVIII** and **XLIX**, is presented for **XXIII** and **XLIV** on p. 132.

AXIOPHILICITY SERIES. Returning to studies that involve good departing groups, DeBruin et al. (*99*) designed a clever experiment in which they varied the departing ability of substituents in competition with the loss of an alkoxy group in analogs of the sterically hindered phosphetanium salt,

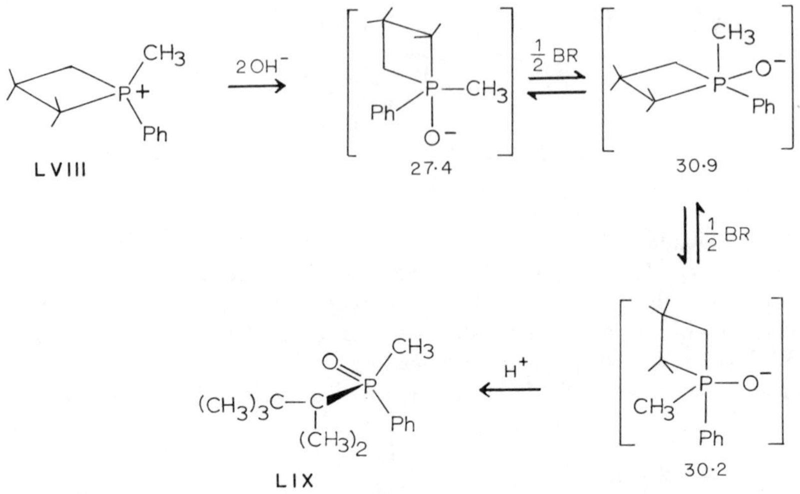

Figure 2.12. A retention mechanism proposed (56) for the hydroxide decomposition of 1-methyl-1-phenyl-2,2,3,3-tetramethylphosphetanium bromide, LVIII. A square pyramidal transition state is included in this variant of the proposed scheme (56) (see p 88).

LVII. The loss of stereospecificity during alkaline hydrolysis in this series (**LX**) is accountable by the multiple pseudorotational process outlined in Figure 2.10. This scheme is written in more detail in Figure 2.13, which summarizes the full cycle of stereochemical relations among reactants and trigonal bipyramidal intermediates, leading to the various phosphetane oxide product formulations encountered in the hydrolysis of 1-X-1-methoxy-2,2,3, 4,4-pentamethylphosphetanium hexachloroantimonates **LX**, studied in 50% aqueous dioxane. The product compositions are in Table 2.9. Similar results were reported (99) for the same series with an ethoxy group in place of the methoxy ligand (Figure 2.14).

Only for salts that contain the $(CH_3)_2N$ substituent is complete retention of configuration observed. This result indicates the poor axiophilicity of the $(CH_3)_2N$ group compared with a methoxy group. For the phosphonium salts that contain either SCH_3 or Cl ligands, loss of stereospecificity is encountered, leading to the formation of isomeric phosphetane oxides lacking these ligands. This is a measure of their greater axiophilicity relative to a OCH_3 group. For the salts containing the OEt and O-iso-Pr ligands, a distribution of isomeric oxides results, either showing loss of OCH_3 groups or loss of one of the other alkoxy ligands contained in the salts. The relative axiophilicity results: $(CH_3)_2N \ll OCH_3 \sim OEt \sim$ O-iso-Pr $\sim SCH_3 <$ Cl. See p 131 for the ordering of axiophilicities of OR groups in an acyclic series.

Although the order does not follow an electronegativity trend, intermediates are traversed that have the OCH_3 group in an axial orientation to account for cis $\rightleftarrows$ trans isomerization between reactant and product when

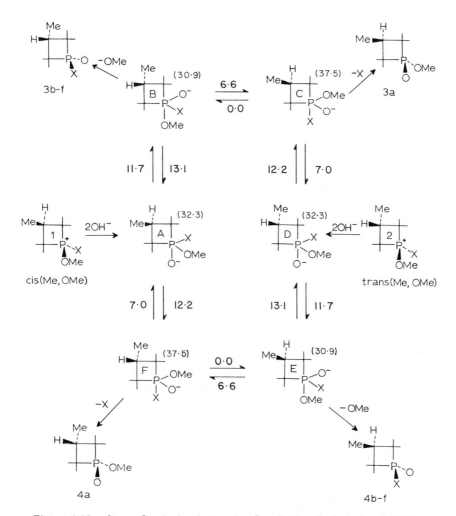

*Figure 2.13. Stereochemical relations in the alkaline hydrolysis of 1-X-1-methoxy-2,2,3,4,4-pentamethylphosphetanium hexachloroantimonates, **LX** (99). Formation of phosphetane oxides via trigonal bipyramidal intermediates undergoing pseudorotation. The numbers in parentheses beside each intermediate are estimated isomer energies (kcal/mol) for X = NMe₂. The numbers along the arrows are estimates of pseudorotational barrier energies (see pp 37 and 88). A steric factor of 2 kcal/mol was applied for the interaction Me₂N–P–OR, as mentioned on p 37. The inversions paths 1 → A → B → 3b-f and 1 → A → F → E → 4b-f were given in Figure 2.3a and Figure 2.4d, respectively, when general mechanisms were discussed.*

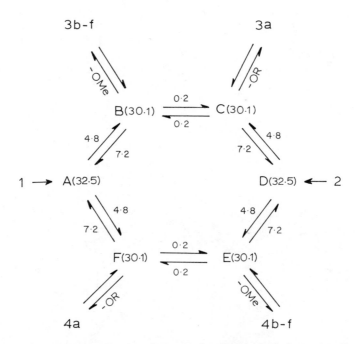

*Figure 2.14.   Postulated intermediates in the alkaline hydrolysis of 1-alkoxy-1-methoxy-2,2,3,4,4-pentamethylphosphetanium salts, **LX**. Estimates of the relative isomer energies of intermediates are shown in parentheses (see p 88). Estimates of pseudorotational barriers are given along the arrows connecting intermediates. Refer to Figure 2.13 for the stereochemical definition of the labels. The almost negligible pseudorotational barriers estimated for B → C and E → F when X = OR compared with the value of 6.6 kcal/mol estimated for these transformations when X = NMe₂ is consistent with complete retention in the latter case relative to loss of stereospecificity for X = OR (cf. Table 2.9).*

the better departing groups, SCH_3 and Cl, are present. These results are similar to those reported on alkaline hydrolysis of acyclic phosphonium salts (*see* Table 2.8 and discussion on p 129) which showed, as here, that alkylthio ligands were better leaving groups than alkoxy ligands. From the observed stereochemistry in the latter series, the initial intermediate is proposed to have an axial orientation of the more electronegative alkoxy group, followed by pseudorotation to bring the alkylthio group to a departing axial position (Figure 2.6). Because of ring constraints in the phosphetanium salts, **LX**—i.e., the axial–equatorial orientation of the four-membered cyclic system —one pseudorotation is required to bring the methoxy group to an axial position.

Table 2.9.   Product Stereochemistry from the Alkaline Hydrolysis of 1-Methoxy-1-X-2,2,3,4,4-pentamethylphosphetanium Salts, LX (99)

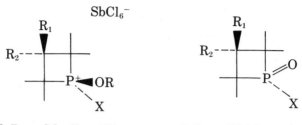

1, R_1 = Me; R_2 = H   3, R_1 = Me; R_2 = H
2, R_1 = H; R_2 = Me   4, R_1 = H; R_2 = Me
   a, X = OMe   d, X = NMe_2
   b, X = OEt   e, X = SMe
   c, X = O-iso-Pr   f, X = Cl

	Phosphetanium salt		*Products (%)*[a,b]			
No.	*X*	*Config*	*3a*	*4a*	*3b–f*	*4b–f*
1a	OMe	—	55	45	—	—
1b	OEt	trans	11	7	49	33
2b	OEt	cis	16	9	39	36
1c	O-iso-Pr	trans	10	4	44	42
2c	O-iso-Pr	cis	15	2	40	43
1d	NMe_2	cis	0	0	100	0
2d	NMe_2	trans	0	0	0	100
1e	SMe	trans	36	61	3	0
2e	SMe	cis	63	37	0	0
1f	Cl	trans	37	63	0	0
2f	Cl	cis	64	36	0	0

[a] Percent total nonacidic products.
[b] Refer to Figure 2.13 for the identification of product stereochemistries.

Journal of the American Chemical Society

To account for the order of axiophilicities encountered in the study of the hydrolysis of the phosphetanium salts, **LX**, DeBruin et al. (*99*) suggested a compromise between substituent electronegativity, OR > $N(CH_3)_2$ ~ Cl > SCH_3, and greater π donor ability of alkoxy compared with Cl and SCH_3 ligands. Presumably, the π donor ability of the $(CH_3)_2N$ group is greater than that for the Cl and SCH_3 ligands as well. This view is consistent with the results of a molecular orbital approach (*100*) and with structural observations (Volume I, p 22), indicating that π donor ligands prefer equatorial orientation in trigonal bipyramids (*see* preference rules on p 88).

In interpreting kinetic axiophilicities, it is important to ascertain the effect of the relative anion stabilization, $R^- + H_2O \rightarrow RH + OH^-$, a factor not explicitly included in the construction of the trigonal bipyramidal preference rules. Also, in this highly hindered series, **LX**, it is difficult to decide whether the usual preference of a large group for an equatorial site of a trigonal bipyramid holds. Hence, steric effects may be entering in the determination of the order of kinetic axiophilicities in a subtle way. As a consequence, it is somewhat misleading to compare orders of axiophilicities obtained on structurally divergent series of compounds.

A further complication is inherent in each of these analyses as usually performed. For five-coordinate phosphorus compounds, activational energies for intramolecular ligand exchange, in general, most probably reflect energy differences between square-pyramid transition states, or some distortion of them, and ground-state trigonal bipyramids. As discussed on p 107, reaction kinetics are likely to be governed by activated states of less-specified geometry, i.e., distorted tetrahedral configurations on the way to five-coordinate intermediates (pseudopentacoordinate structures). For the latter, a parallel is assumed between the associated barrier energies and relative energies of five-coordinate intermediates. The relation may be viewed as an increasingly better approximation for rate comparisons within a series than between different series.

With the above caveats in mind, the comparison of numerical estimates of isomer energies of proposed intermediates in the hydrolysis of the Phosphetanium Salts **LX** (X = NM$_2$, Figure 2.13, and X = OR, Figure 2.14) tend to support the product stereochemistries in Table 2.9. High pseudorotational barriers are predicted for X = NMe$_2$ relative to the alkoxy derivatives (X = OR). In view of the greater leaving ability of the methoxy group compared with that of the NMe$_2$ group (low barrier from **B** to **3d** in Figure 2.13), it is reasonable that exclusive formation of **3d** from **ld** and of **4d** from **2d** should occur. For any phosphetane oxide of opposite configuration (**ld** → **4d** or **2d** → **3d**), to form, additional high-energy pseudorotations are necessary (**A** → **F** → **E** for **ld** and **D** → **C** → **B** for **2d**).

For the alkoxy derivatives, low barriers (which are more comparable in energy) are estimated—e.g., for **A** → **B** and **A** → **F** → **E** (for **lb** and **lc** of Figure 2.14). Thus, formation of **3b-c** and **4b-c** are more nearly competitive. The leaving ability of the OMe group is better than that of the higher alkoxy groups but not greatly so; thus, the formation of some **3a** from **B** and **4a** from **F** is rationalized.

Although detailed calibrating data are unavailable to allow similar estimates of isomer energies for phosphetanes, **LX**, containing the SMe or Cl groups, these derivatives most likely follow an energy profile analogous to that for the alkoxy derivatives (Figure 2.14), resulting in rapid pseudorotation among the Intermediates **A** to **F** but modified to account for the superior leaving ability of the SMe and Cl groups—i.e., lower barriers for the processes: **C** → **3a** and **F** → **4a**.

It is worthwhile to examine additional axiophilicities to see how they compare among each other. From a study of NMR spectra on five-coordinate spirophosphoranes, Trippett and co-workers (*101*) conclude that the OPh and SPh have similar axiophilicities. Gorenstein (*30*) showed little difference between the axiophilicity for an OEt and OPh group obtained from NMR data on monocyclic alkyloxyphosphoranes. In an NMR study of 1,3,2-dioxaphospholanes (**LXI**), Oram and Trippett (*102*) found the axiophilicic order: OPh > $(CH_3)_2N \sim CH_3 \gtrless$ iso-Pr > Ph. These authors stress the extent of π donor ability of equatorial nitrogen to account for its low axiophilicity,

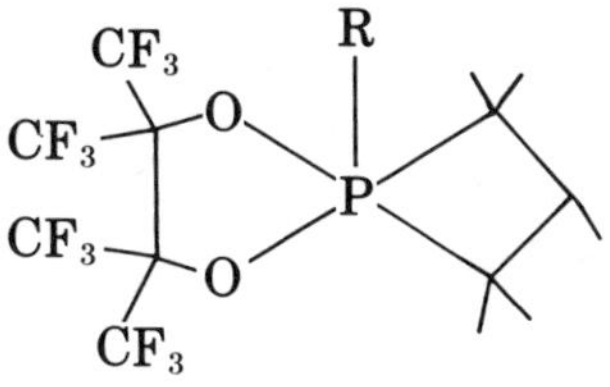

| **LXI** trans | and | **LXI** cis |

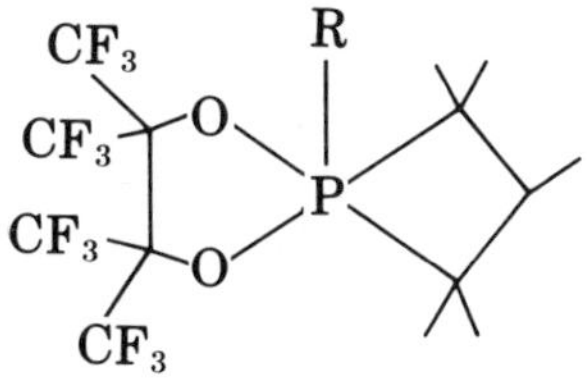

near that of the iso-Pr group. The latter group is comparable in size with the $(CH_3)_2N$ group but of considerably lower electronegativity.

An X-ray study of **LXI** (R = p-BrC$_6$H$_4$) (*103*) has shown that the ground-state structure (Volume I, p 20) is a basic square pyramid. Assuming that this structure prevails in solution, the $\Delta G^{\ddagger}$ values from the NMR study on derivatives of **LXI** should reflect the energy required to place the R group at an axial site in a trigonal bipyramidal transition state (cf. Volume I, Figure 3.11b) relative to its location at an apical site in the square pyramid (see

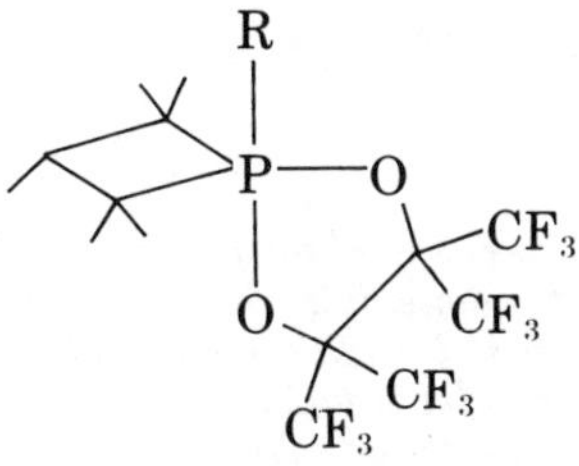

Table 1.11, Entries **32–40**, p 46). However, it is argued (*103*) that the energy difference between the ground-state square pyramid and the usually observed ground-state trigonal bipyramid is small compared with the activation energies of 9–22 kcal/mol obtained in this series.

Marsi's studies (*104, 105, 106, 107*) on the alkaline cleavage of phospholanium salts involving a five-membered ring (**LXII**) supported the axiophilicity series: OSiCl$_3$ > OCH$_3$ > CH$_2$Ph > Ph. Earlier studies on the

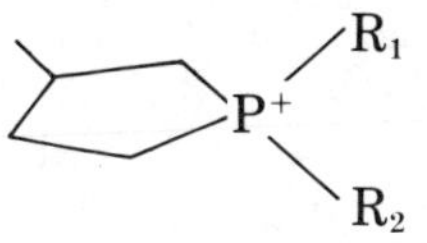

LXII

product ratios from the decomposition of acyclic phosphonium hydroxides led to the following seqeunce for the ease of elimination of various groups:

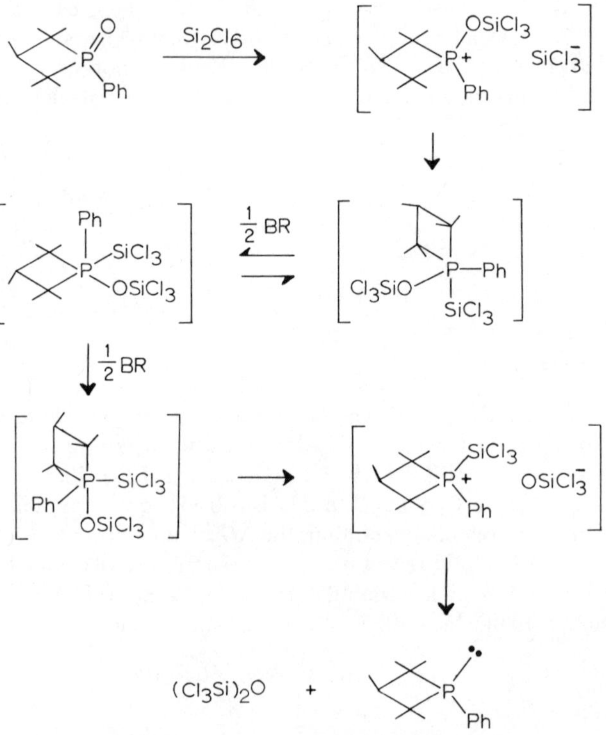

Figure 2.15. Proposed mechanism for the stereospecific reduction of cis- *or* trans-1-phenyl-2,2,3,4,4--pentamethylphosphetane 1-oxide *(trans illustrated) with* Si_2Cl_6 *to give the respective phosphetane with retention of configuration (95, 112). A square pyramidal transition state is included in this variant of the proposed scheme (see p 87).*

$CH_2Ph > Ph > CH_3 > Et >$ higher alkyls (*108, 109*). Cavell and co-workers (*110, 111*) established ground-state structures for a series of acyclic phosphoranes by observing characteristic coupling constants for equatorial and axial substitution. This led to the axiophilicity series $F > Cl$, $Br > CF_3 > OR$, SR, NR_2, R.

Surprisingly, no great differences in all of these series exist where it is possible to compare overlap, other than the placement of the phenyl group by Oram and Trippett (*102*) at the low end of the axiophilicity scale. Because of the greater uncertainties inherent in their study regarding assessment of the ground-state structures (as noted above), the result on Ph may be spurious. Thus, it is possible to construct a general ordering of axiophilicities applicable to acyclic and cyclic derivatives: $F > OSiCl_3 \gtrsim Cl (> CF_3) \sim SCH_3 > SPh \sim OPh \sim$ O-iso-Pr $\sim OEt \sim OCH_3 > CH_2Ph > Ph > (CH_3)_2N \sim CH_3 \gtrsim$ iso-Pr > higher alkyls. The ordering of the trichlorosiloxy group is more approximate than that for the others. The series shows a general

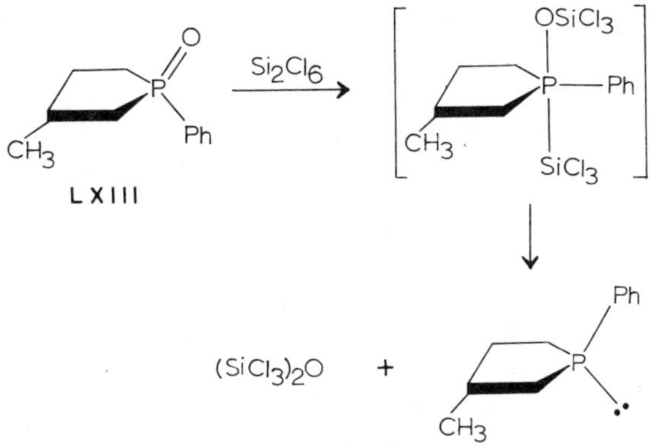

Figure 2.16. Inversion of configuration in the deoxygenation of 3-methyl-1-phenyl-phospholane 1-oxide, LXIII, by Si₂Cl₆ (106, 117)

correlation with the electronegativity scale given in Table 2.1. Pi bonding effects provide a secondary contribution to group axiophilicity which most likely follows the order: $R_2N > RO > RS$.

Comparisons Among Four-, Five-, and Six-Membered Ring Systems. LIGAND AXIOPHILICITY. For phosphetane oxides containing departing groups at the upper end of the axiophilicity scale, reactions invariably proceed by a retention pathway (Figure 2.1) and show little evidence of isomer cross-over (*81, 82, 112–115*). Some examples are on p 133 and in Figure 2.7. A related mechanism is postulated for the deoxygenation of 1-phenyl-2,2,3,4,4-pentamethylphosphetane 1-oxide with Si_2Cl_6 (*95, 112*) (Figure 2.15). Both the attacking nucleophile (the trichlorosilyl anion), and the departing group (the trichlorosiloxide ion) are relatively electronegative. This contrasts with the inversion process that takes place in the corresponding deoxygenations of acyclic chiral phosphine oxides (*116*) and phospholane oxides (*106,117, 118*) (five-membered ring systems). The different stereochemistries encountered in these systems are associated with the unfavorable energy required to place the four-membered ring in diequatorial positions in the trigonal bipyramidal intermediate. The deoxygenation of 3-methyl-1-phenyl-phospholane-1-oxide, **LXIII**, by Si_2Cl_6, reported by Mislow and co-workers (*106,117,118*), is illustrated in Figure 2.16.

REACTION RATES AND STEREOSPECIFICITY. Although location of a ring diequatorially in a trigonal bipyramidal intermediate is not postulated for reactions of four-membered cyclic systems (and only when highly axiophilic substituents are present for reactions with five-membered cyclic structures), the formation of such intermediates should be a common occurrence for derivatives containing six-membered rings. Consequently, an inversion mechanism should dominate similar to that generally observed for nucleophilic reactions of acyclic compounds (*see* p 116). However, when acyclic derivatives possess unusual steric effects (p 119) or involve the competitive departure of groups of similar axiophilicity (Figure 2.6), a retention mechanism

Table 2.10. Rates and Activation Parameters for the

Entry   *Compound*

	$R_1{}^a$	R_2	Rate Constant $(L^2\ mol^{-2}\ sec^{-1b})$
			$15°$
1	CH_2Ph	Ph	1.12×10^3
2[d]	Ph	CH_3	1.24

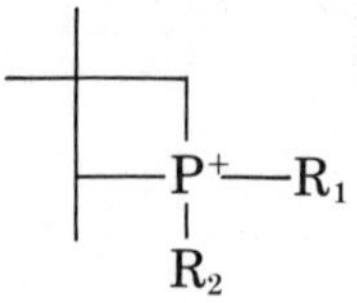

3	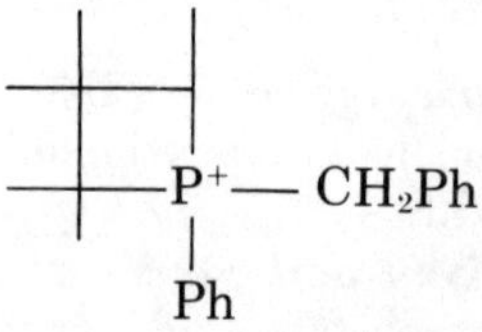 P^+—CH_2Ph, Ph		70.7

	$R_1{}^a$	R_2	
4	CH_2Ph	Ph	40.7
5	CH_2Ph	CH_2Ph	1.69
6	CH_2Ph	CH_3	0.413
7[e]	Ph	CH_3	5.93×10^{-2}

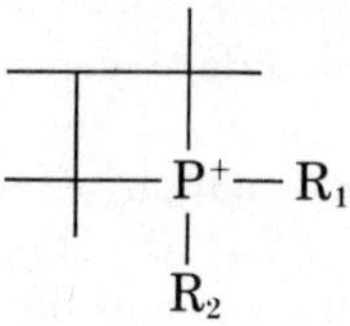

8	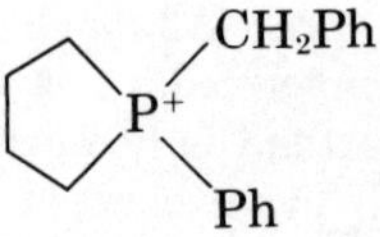	5.35×10^{-2}

9	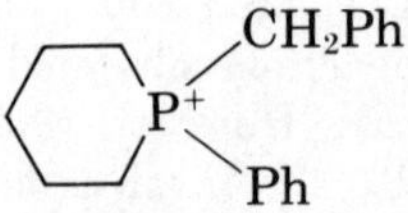	3.86×10^{-3} (50°C)

Alkaline Decomposition of Cyclic Phosphonium Salts (*56*)

Rate Constant $(L^2 mol^{-2} sec^{-1b})$ 25°C	$\Delta H^{\neq}$ *(kcal/mol)*	$\Delta S^{\neq c}$ *(eu)*
2.77×10^3	16.0	10.9
3.44	17.3	1.9
165	15.0	2.0
111	18.0	11.0
5.56	18.8	7.8
1.37	21.2	13.2
0.214	20.3	6.5
0.171	18.2	−1.6
5.6×10^{-4}	14.0	−26.4

Table 2.10.

$R_1{}^a$	R_2	Rate Constant $(L^2\ mol^{-1}\ sec^{-1b})$
		$15°$
10		

$$
\begin{array}{c}
CH_3 \quad\quad CH_2Ph \\
\diagdown \quad\diagup \\
P^+ \\
\diagup \quad\diagdown \\
C_2H_5 \quad\quad\; Ph
\end{array}
$$

    4.75×10^{-3} (50°C)

[a] R_1 is the departing group unless otherwise specified.
[b] Third-order rate constants. Total ionic strength equals $0.1M$ in 50% ethanol-water.

is possible. Analogously, the presence of steric effects in six-membered ring structures may cause some reactions to lack stereospecificity.

In connection with comparisons among cyclic structures varying in ring size, rate data obtained by Cremer and co-workers (56) for the alkaline decomposition of some cyclic phosphonium salts are summarized in Table 2.10. As discussed (p 141), rate enhancement for the alkaline hydrolysis of four-membered rings correlates with a reduction in the number of α-methyl ring substituents. This same rate effect holds for the entries in Table 2.10 (**1 > 3 > 4**). The rate increases, as expected, for the compound containing the better departing group (**1 > 2, 4 > 7**). The rate decreases with increasing ring size (**1 > 8 > 9**). Also, the rate is influenced by the nondeparting groups (**4 > 5 > 6**, see p 123).

A comparison of initial intermediates postulated in the case of Entries **1, 8,** and **9** is shown in Figure 2.17. The strain energy relief on forming these intermediates is higher in the smaller ring containing salts. Even though strain energy may be relatively unchanged or actually increased on formation of the Intermediate **C**, the location of the benzyl group axially is likely to enhance its formation. In this regard, the low enthalpy of activation reported for the six-membered ring structure (Entry **9**) may reflect a changeover in mechanism.

The favorable entropy factor for Compounds **1** and **8** relative to **9** is an indication of the presence of less rigid intermediates, **A** and **B** (Figure 2.17). The accompanying larger degree of motional freedom is suggestive of pseudorotation (p 104). Thus, a retention mechanism for the four- and five-membered cyclic phosphonium salts, in contrast to a direct displacement of the benzyl group for the six-membered cyclic derivative, provides a rational basis upon which to operate.

Additional insight concerning this apparent mechanistic variation with ring size is obtained from the examination of the stereochemical course of the hydrolysis reaction of chiral derivatives. Results based largely on studies

(Continued)

Rate Constant $(L^2\ mol^{-1}\ sec^{-1b})$ 25°C	$\Delta H^{\neq}$ (kcal/mol)	$\Delta S^{\neq c}$ (eu)
3.29×10^{-4}	20.2	-6.8

[c] Values of $\Delta S^{\neq}$ refer to 25°C.
[d] Ring opening takes place.
[e] Ring expansion takes place.

Journal of Organic Chemistry

by Marsi and co-workers (104–107, 119–122) are listed in Table 2.11. By examining the product stereochemistry and noting the footnotes to Table 2.11, competition between stereoelectronic and ring strain effects is apparent. A reduction of ring strain as ring size increases is evident from the stereochemical outcome for reaction of the various cyclic structures containing both the phenyl group and departing benzyl group. Complete retention for the five-membered cyclic salt shifts to a product mixture for the six-membered ring compound and finally to complete inversion for the seven-membered derivative. Again, in the presence of a better departing ligand (methoxy), 100% inversion is observed for the six-membered cyclic compound compared with a product mixture for the salt that contains the five-membered ring.

These results support the presence of the two basic mechanisms outlined in Figure 2.1, the retention route involving one pseudorotation and the inversion pathway with the ring systems located diequatorially. For the Entries (8–9) and (10–11), both mechanisms seem to operate in competition with each other. Apparently, a balance between ring strain, axiophilic tendencies, and steric strain (the latter resulting from incorporation of the ring methyl substituent) is achieved for these substances.

Marsi (104) alluded to a steric effect between the ring methyl and attacking hydroxide ion as the possible cause of the reduced amount of inver-

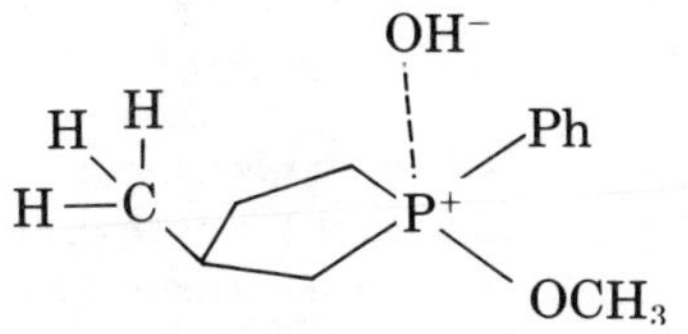

trans (9) 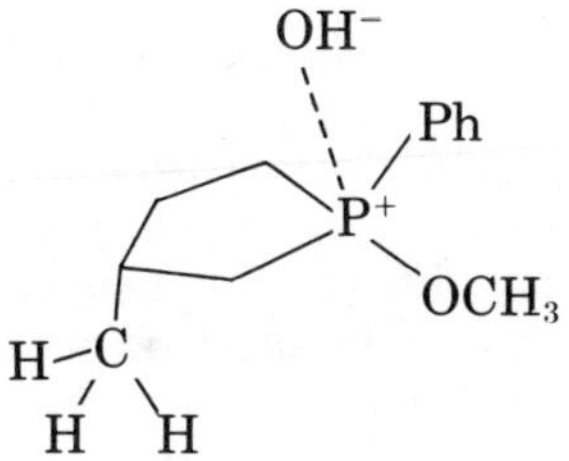   cis (8)

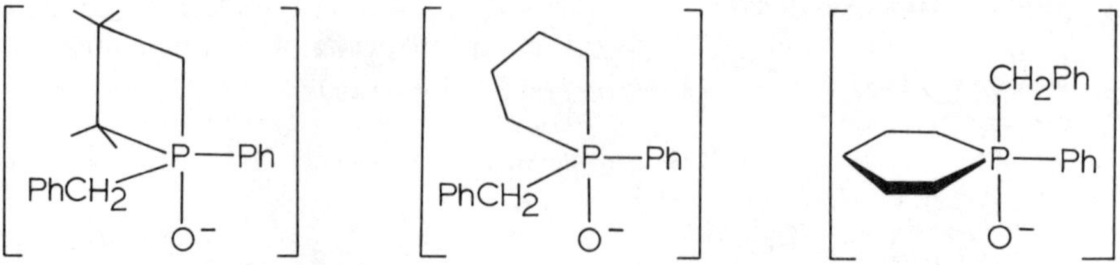

*Figure 2.17. Postulated Intermediates **A**, **B**, and **C** in the alkaline decomposition of the respective Entries **1, 8**, and **9** of Table 2.10*

sion at phosphorus for the trans (**9**) compared with the cis (**8**) methoxyphospholanium derivatives listed in Table 2.11. In the phosphorinanium salts, (**10** and **11**), the reduction in inversion for the trans relative to the cis isomer is even greater (*121*). A similar steric argument may be extended to these derivatives. For direct inversion to occur, the hydroxide must approach opposite the departing benzyl group. We have simulated the hydrolysis reaction using the molecular mechanics approach outlined for ribonuclease action (p 199ff). The reaction coordinate for the trans isomer shows a repulsion between the 4-methyl group and the benzyl group that leads to a high-energy transition state. In the cis isomer, the positioning of the 4-methyl group away from the benzyl group removes the steric interaction between them. In agree-

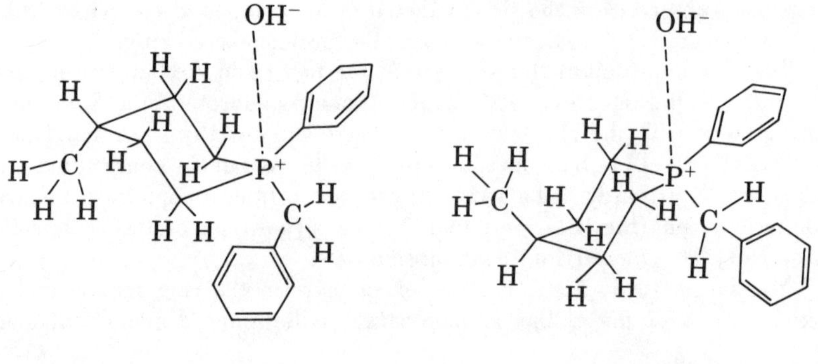

trans (**11**)   cis (**10**)

ment, simulation of the pathway for the cis isomer shows a relatively low energy profile for hydroxide attack. In either isomer, the 4-methyl group by itself is located too far from the phosphorus center to produce any appreciable steric effect with the attacking nucleophile.

Only in the cis- and trans-phospholanium derivatives (Entries **5** and **6**) containing poor leaving groups is there evidence (*see* Footnote *d* of Table 2.11) suggesting stereomutation of phosphorus—i.e., the appearance of a product mixture of diastereomeric phosphine oxides. The closeness in axiophilicity of the methyl and phenyl groups indicates low pseudorotational barriers. Thus, the lack of stereospecificity is accounted for by successive pseudorotations which equilibrate the reactant isomers. This process is similar to that shown in Figure 2.13 for the stereomutation of phosphetanium salts undergoing alkaline hydrolysis.

There is no evidence from the preceding mechanistic considerations to support diequatorial placement of a four-membered ring in a trigonal bipyramidal intermediate, even though the departing groups have high axiophilicities, i.e., $OSiCl_3$ (Figure 2.15). In the case of five-membered rings, diequatorial placement is indicated when the departing axial group is OCH_3 (p 153) but not when it is phenyl (p 154). When six-membered rings are present, they seem to enter diequatorial positions even for intermediates containing substituents of low axiophilic character, such as the phenyl and benzyl groups (p 117 and p 154). All of the studies that have been discussed involved carbon ring atoms directly bonded to phosphorus.

If the directly attached ring atoms were of higher electronegativity, greater difficulty would be expected in forming intermediate trigonal bipyramids with diequatorial ring placement because of the reduced axiophilicity difference between the departing ligand and the hetero ring atom. Consequently, the studies involving the carbon-containing cyclic structures are the most favorable to equatorial ring placement and, therefore, represent limiting cases for the respective ring size examined.

UNSATURATED RINGS. With unsaturated carbon-containing ring systems, the tendency for ring opening and ring expansion increases over that for saturated counterparts. This observation has been taken as an indication of a greater degree of ring strain in the trigonal bipyramidal intermediate containing the unsaturated ring (*125*). (Both electron delocalization into the ring from the more loosely held electrons in the axial P–C ring bond and a ring strain effect due to a less than "perfect" fit of the planar unsaturated five-membered ring to an equatorial–axial orientation in the trigonal bypyramidal intermediate may contribute to increased ring fission for unsaturated ring systems.) Comparison of bond distance variations between saturated and unsaturated five-membered rings in phosphorane structures also supports the view that increased strain exists in unsaturated cyclic systems (*6, 7, 126*) (cf. Volume I, p 46).

As an example, the hydrolysis of Phosphonium Salt LXIV proceeds with ring opening to give Phosphine Oxide LXV (*127, 128, 129, 130, 131*) (Figure 2.18). Reaction is postulated to involve direct ring cleavage via a trigonal bipyramidal intermediate that orients the five-membered ring in the usual axial–equatorial configuration. The energy of activation for this ring-opening reaction is 18.9 kcal/mol (R = CH_3) (*129, 130*). For the alkaline hydrolysis of the related saturated cyclic phosphonium salt, XXVI (p 115), which occurred with ring retention, a much higher energy of activation was reported (37.6 kcal/mol) (*49*). Third-order kinetics implicated the formation of the product phosphine oxide with loss of a phenyl group from the pseudorotated intermediate as the slow step. With better leaving groups, R = NEt_2 or

Table 2.11.   Product Stereochemistry in the Alkaline Cleavage of Phospholanium, Phosphorinanium, and Phosphepanium Salts[a]

Entry   Compound   Product[b]

		R_1	R_2	Retention	Inversion	Ref
1	cis	CH_2Ph	CH_3	100	0	105, 106, 107, 119, 120
2	trans	CH_2Ph	CH_3	100	0	
3	cis	CH_2Ph	Ph	100	0[c]	106, 107, 120
4	trans	CH_2Ph	Ph	100	0	
5	cis	Ph	CH_3	50	50[d]	107, 120
6	trans	Ph	CH_3	50	50	
7		Ph	Ph	—	—	
8	cis	OCH_3	Ph	42	58[e,f]	104
9	trans	OCH_3	Ph	51	49	

[a] Cis and trans refer to the orientation of R_2 and the ring methyl group except when the methoxy group is present. There cis and trans refer to the methoxy group (R_1) and the ring methyl. Retention refers to reactant–product relations, cis–cis and trans–trans, while inversion refers to the relations, cis–trans and trans–cis.

[b] The phosphine oxides are configurationally stable toward aqueous hydroxide. See Ref 54, 105–107, 119, and 121.

[c] An earlier suggestion (105) that retention might result from equatorial loss of benzyl because pseudorotation should produce both retained and inverted products apparently was abandoned and replaced by the mechanism discussed here (104, 123).

[d] In view of Footnote b, a common intermediate preceding phosphine oxide formation is indicated. Analysis of cis and trans Isomers 5 and 6, which have been treated separately with hydroxide under cleavage conditions and allowed to partially

CH_2Ph in **LXIV**, hydrolysis gives the oxide shown, with retention of the five-membered ring (131.).

Both of the processes, loss of a phenyl group and ring opening, occur on alkaline hydrolysis of the phospholenium salt, **LXVI** (Figure 2.19). The formation of the major product (**LXVII**) by loss of the phenyl anion, in spite of the greater stability of the allyl system in **LXVIII**, has been partially attributed to steric effects between the ring methyl groups, which are postulated

Table 2.11. Continued

Entry	Compound	Product[b]

		R_1	R_2	Reten-tion	Inver-sion	Ref
10	cis	CH_2Ph	Ph	48	52[f,g]	*121*
11	trans	CH_2Ph	Ph	78	22	
12	cis	OCH_3	Ph	0	100	*104*
13	trans	OCH_3	Ph	0	100	

| **14** | cis | CH_2Ph | Ph | 0 | 100 | *122* |
| **15** | trans | CH_2Ph | Ph | 0 | 100 | |

react, gives approximate 1:1 mixture of cis and trans salts for the remaining un-decomposed salts (*107*). The mechanism in this case may be analogous to that out-lined in Figure 2.10.

e Of the retained product for each reactant isomer, 11% results from attack of hydroxide at the methoxy carbon atom, as deduced from ^{18}O labeling of the methoxy substituent (*104*). The cis and trans stereochemistry of the Reactants **8** and **9** was established indirectly by a single crystal, X-ray study (*124*).

f A common intermediate does not result in these cases since the cis and trans isomers give different ratios of the same products, even when the percent retention is corrected for hydroxide attack at the methoxy carbon.

g As in (*e*), 9 % of the retained product results from hydroxide attack at the methoxy carbon (*104*).

to lead to poor overlap of the allylic pi system with the *p* orbital on the incipient carbanion (*65*). This loss of resonance stabilization reduces the effectiveness of the ring-opening path in competition to the departure of the phenyl anion.

Other phospholenium salts containing five-membered rings follow the above mechanistic routes. Ring expansion occurs in the alkaline hydrolysis of Phosphonium Salt **LXIX** (*131*) according to a scheme simlar to that in Figure 2.18. The postulated intermediate formed on ring opening suffers the loss of iodide and leads to the product Oxide, **LXX**, with ring expansion (Figure 2.20). If iodine is substituted by OCH_3 in **LXIX**, alkaline hydrolysis proceeds exclusively with ring opening to give Phosphine Oxide **LXV**, R = CH_3OCH_2 (Figure 2.18) (*131*). Ring expansion presumably does not occur because the methoxide ion is a poorer leaving group compared with iodide.

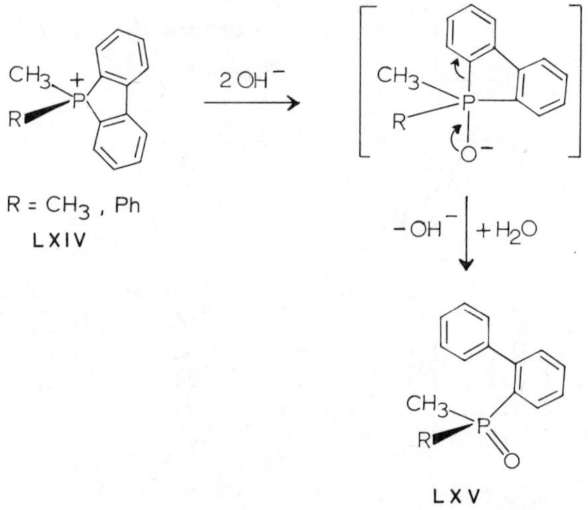

Figure 2.18. Mechanism of alkaline hydrolysis of dibenzophosphonium salts **(LXIV)** *undergoing ring cleavage to give biphenyl-2-ylphosphine oxides* **(LXV)**. *A third-order rate law is followed, which suggests that collapse of the intermediate is rate determining (129, 130).*

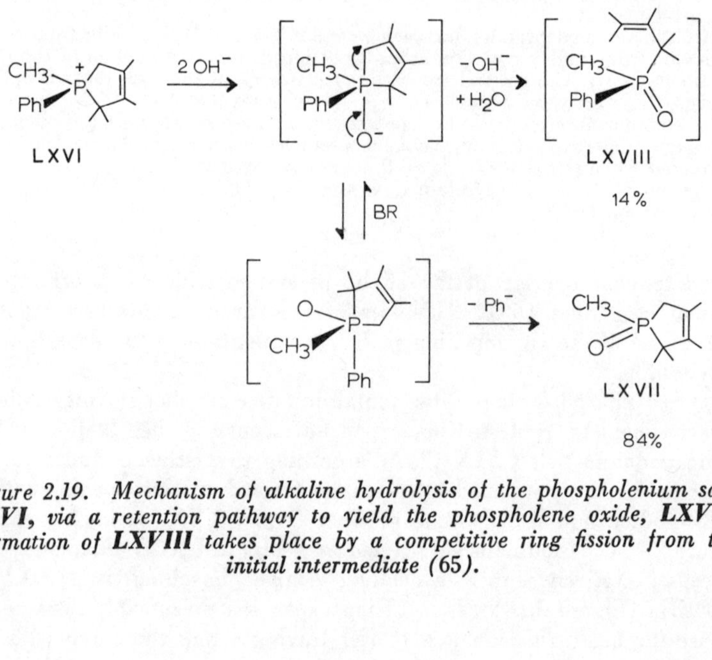

Figure 2.19. Mechanism of alkaline hydrolysis of the phospholenium salt, **LXVI,** *via a retention pathway to yield the phospholene oxide,* **LXVII.** *Formation of* **LXVIII** *takes place by a competitive ring fission from the initial intermediate (65).*

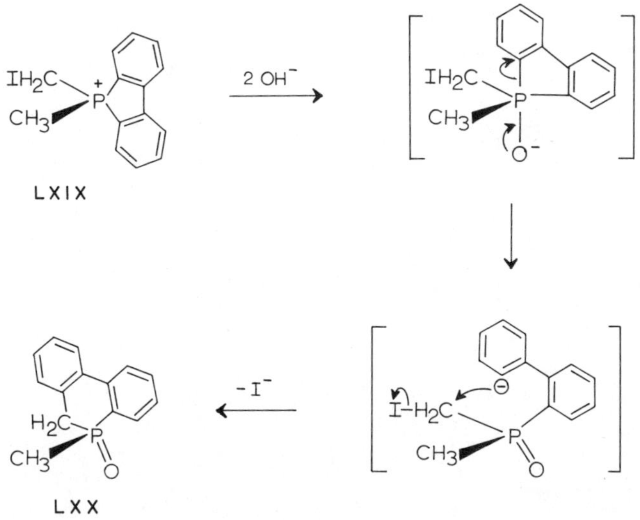

*Figure 2.20. Ring expansion in the alkaline hydrolysis of 9-iodomethyl-9-methyl-9-phosphoniafluorene iodide (**LXIX**) to give 9-methyl-9, 10-dihydro-9-phosphaphenanthrene-9-oxide (**LXX**) (cf. Figure 2.9). Substitution of phenyl in place of methyl produces similar results (131).*

Both Phosphonium Salts **LXXI** (*132, 133*) and **LXXII** (*134*) undergo ring opening during alkaline hydrolysis while **LXXI** also forms phospholene oxide with loss of the phenyl group. The amount of the latter oxide formed depends dramatically on the reaction conditions. The activation energy for ring fission in **LXXII** is only 11 kcal/mol for R = H and 12.2 kcal/mol when R = Ph (*134*). These low barriers again indicate the increased ring strain present in unsaturated ring systems.

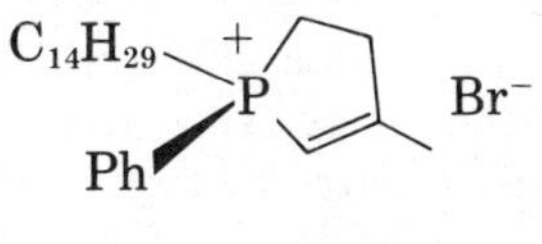

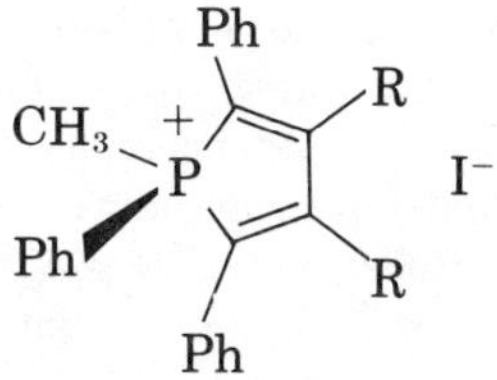

LXXIa **LXXII** R = H, Ph

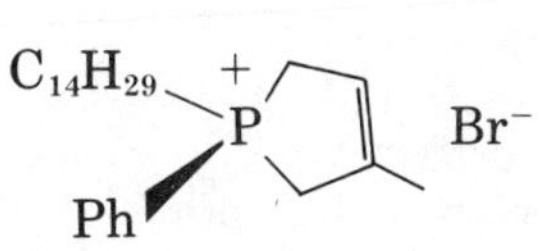

LXXIb

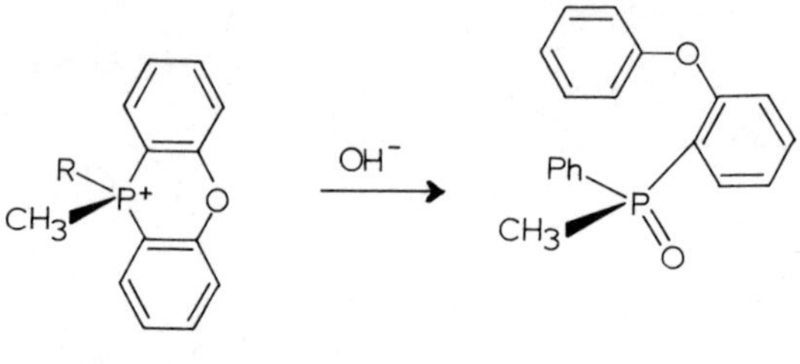

R = CH₃, Ph

LXXIII

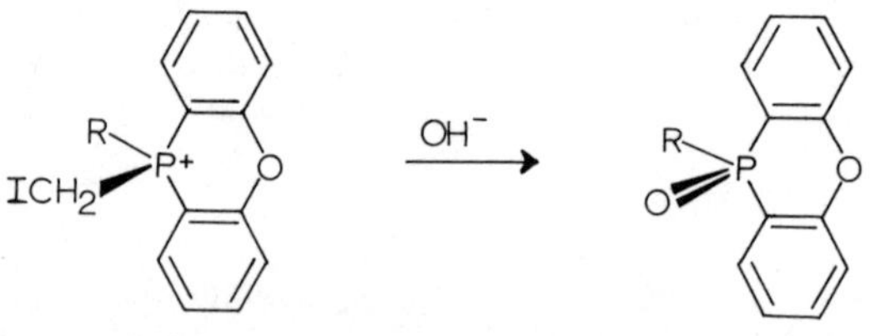

R = CH₃, Ph

LXXIV

Figure 2.21. Ring opening and ring retention in the alkaline hydrolysis of the respective phenoxaphosphonium salts, **LXXIII** *and* **LXXIV** *(129, 131)*

Similar results are obtained from studies of the alkaline hydrolysis of unsaturated six-membered cyclic phosphonium salts *(129, 130, 131)*. Ring fission occurs for **LXXIII** *(129, 131)* while departure of CH_2I from **LXXIV** *(131)* results in ring preservation in the product phosphine oxide (Figure 2.21).

The ring systems seem to be essentially planar. For example, NMR data *(131)* show magnetic equivalence of the methyl groups down to $-55°C$ for **LXXIII** (R = CH_3). The accompanying ring strain effect, which presumably is less than that for an unsaturated five-membered ring, still appears to make itself evident. Relief of ring strain is best achieved by the preferential location of the ring at an axial–equatorial site in the intermediate trigonal bipyramid. In saturated six-membered cyclic phosphonium salts containing ring substituents (p 153), both the latter ring placement as well as a diequatorial disposition have been postulated to take part in competitive processes.

With axial–equatorial placement, the phosphonium salt, **LXXIII**, is postulated *(129, 131)* to hydrolyze by the route shown in Figure 2.18, whereas, the initial pentacoordinated intermediate formed from **LXXIV** pseudorotates to bring the CH_2I group to a leaving site before oxide formation occurs. Presumably oxide formation takes place with **LXXIV** and not **LXXIII** because CH_2I is a better leaving group than CH_3. The hydrolysis of **LXXIII**

($R = CH_3$) is a third-order process, as is that for **LXIV** (Figure 2.18). Analysis yields an energy of activation of 21.8 kcal/mol (*129*). This value is only slightly higher than the value (18.9 kcal/mol) determined for the hydrolysis of **LXIV** ($R = CH_3$).

Examples of ring expansion, like that observed for four- and five-membered compounds, do not seem to have been reported for six-membered ring derivatives whether the rings are unsaturated or saturated (cf. pp. 134, 159).

BISPHOSPHONIUM CENTERS AND BICYCLICS. Several nucleophilic displacement reactions have been studied with bisphosphonium salts containing six-membered and larger rings and bicyclic phosphonium compounds containing a variety of ring sizes (*135*). The primary new feature obtained with the bisphosphonium salts is the high rate of alkaline hydrolysis compared with that of acyclic mono- and acyclic bisphosphonium salts (*136*). This effect has been attributed to relief of electrostatic repulsion between the two phosphonium centers, for example, that which occurs on formation of the intermediate monophosphonium salt, **LXXVI**, from the bis salt, **LXXV** (*136*).

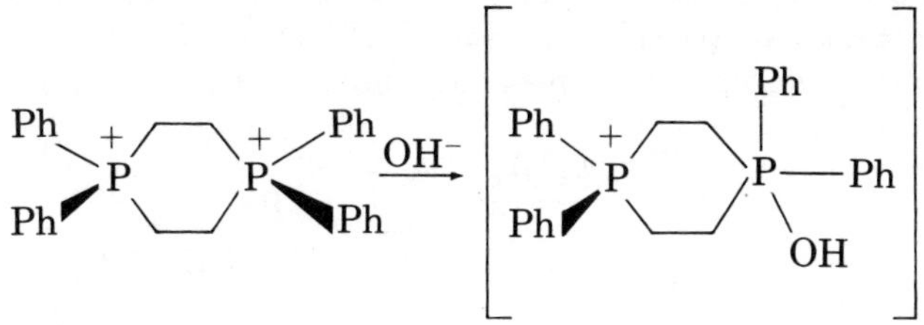

LXXV   **LXXVI**

With excess hydroxide ion the products differ from that obtained when the bis salt is in excess (*136*).

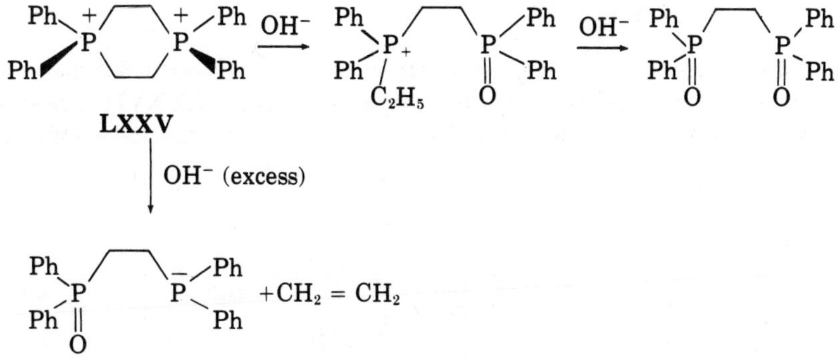

LXXV

If a better departing ligand (the benzyl group) is substituted for a phenyl group in **LXXV**, one at each phosphorus center, alkaline hydrolysis proceeds with exocyclic cleavage to give a mixture of isomeric oxides (*137*).

The mechanism here may be similar to that discussed on p 153 for partial inversion reported in the case of related phosphorinanium salts.

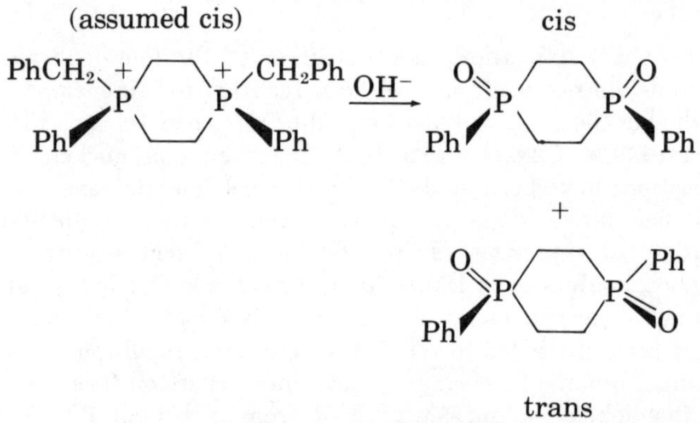

(assumed cis)   cis

+

trans

An interesting rate acceleration was reported for the alkaline decomposition of the spirocyclic phosphonium salt, **LXXVII**, in 95% aqueous ethanol. The hydrolysis rate was faster than that observed for ring cleavage in the

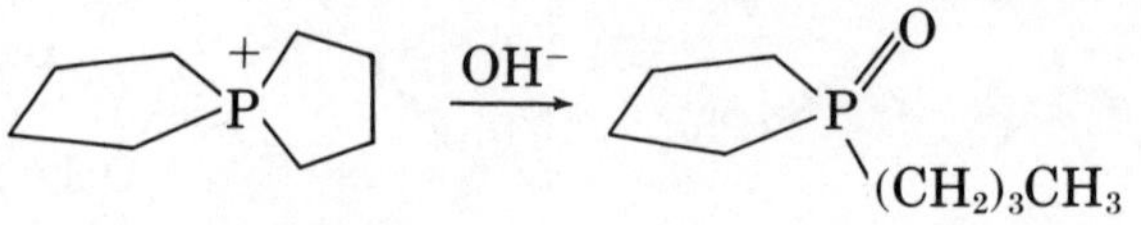

LXXVII

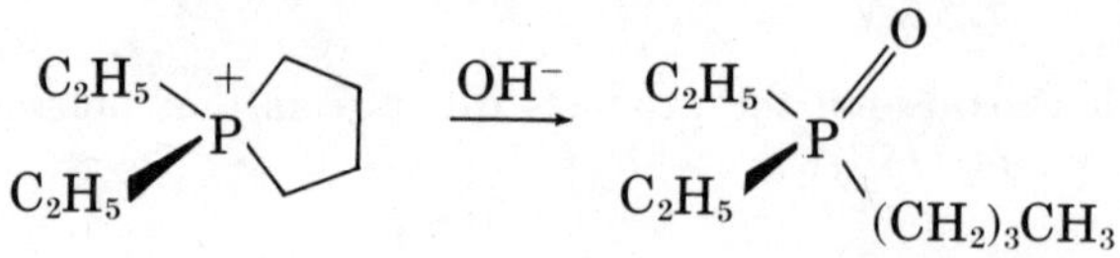

LXXVIII

related monocyclic derivative, **LXXVIII** (138). It is expected that the relief of ring strain in **LXXVIII** would be greater than that for **LXXVII** if reaction proceeds through trigonal-bipyramidal intermediates of the following type.

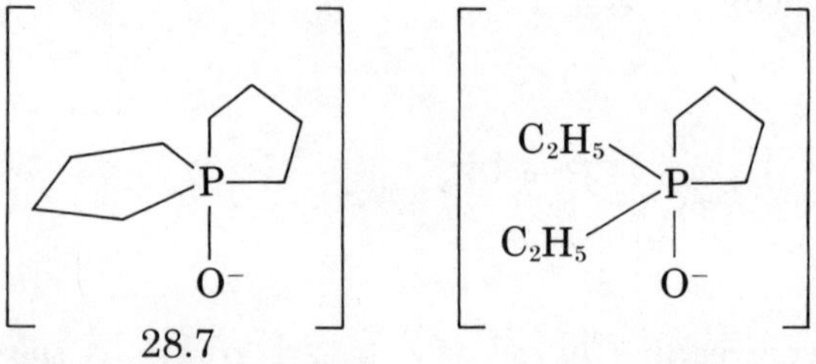

28.7

From **LXXVII**   From **LXXVIII**

As pointed out on p 107, the importance of square pyramidal over trigonal bipyramidal intermediates arises for spirocyclic compounds. Accordingly, formation of the square pyramidal intermediate, **LXXIX**, would result in the simultaneous relief of strain in both rings of the spirocyclic salt, **LXXVII**. Further study, however, is necessary before a satisfactory mechanism is established.

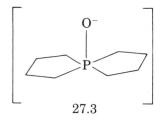

27.3

LXXIX

Nucleophilic Substitution at Tricoordinated Phosphorus

Model Pseudopentacoordinate Structures. An area in which mechanistic postulations are subject to considerable uncertainty concerns reactions of tricoordinated phosphorus. For these substances, addition of an attacking group leads to a pseudopentacoordinated structure if the nonbonding electron pair on phosphorus is assumed to occupy a coordination site.

Suitable models for intermediates of this kind are somewhat rare (*38*), especially those with adequate structural characterization. SF_4, which possesses a structure closely resembling OSF_4 (*see* Volume I, Table 2.1), is perhaps the best example. The equatorial F–S–F angle in SF_4 is 101°. In OSF_4 this angle widens to 110°. It has been argued (Volume I, p 11) that the narrow angle as well as the rest of the structural parameters are a consequence of the greater repulsion associated with nonbonding electron pairs compared with bonding electron pairs. A number of other examples, cyclic sulfuranes, have similar structures (*3, 139, 140*). Nucleophilic substitution on small-membered, cyclic tervalent phosphorus compounds, presumably having ring strain similar to their tetracoordinated counterparts, can proceed through intermediates appropriately modeled after such distored pseudotrigonal bipyramids (*38*).

Hudson and Brown (*38*) have pointed out that for these reactions less of an energy difference is expected between the reaction intermediates, **LXXX** and **LXXXI**, than for corresponding intermediates in reactions of related tetracoordinated compounds. On this basis, an inversion path becomes a more energetically accessible path for pseudopentacoordinated structures (*3*). In effect, this concept results in reversal of Rules 1 and 2 (p 88).

Four-Membered Cyclic Phosphetanes. An example where this scheme seems to be followed is provided by 1-chloro-2,2,3,4,4-pentamethylphosphetane, **LXXXII**, which undergoes nucleophilic substitution (*141, 142*) with predominant inversion of configuration (*141*). This observation contrasts with the retention observed (*48, 82, 90*) in similar substitutions on

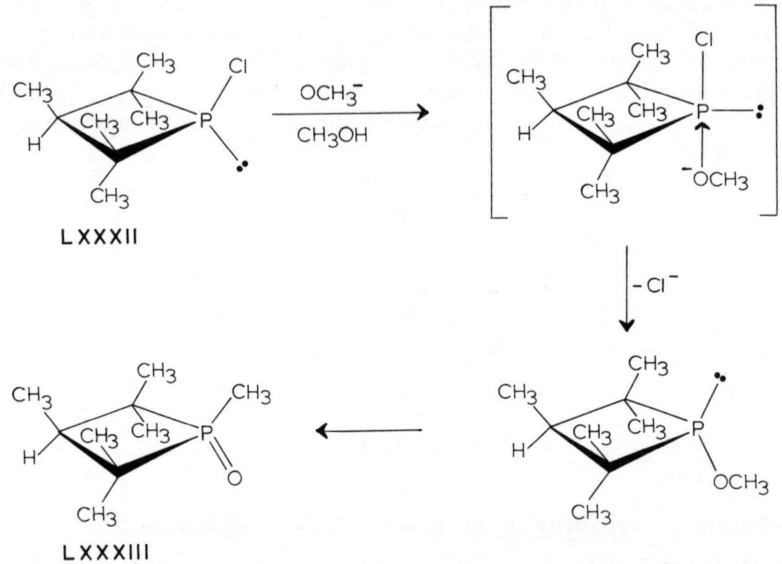

*Figure 2.22. Proposed mechanism for the reaction of 1-chloro-2,2,3,4,4-pentamethylphosphetane, **LXXXII**, with sodium methoxide in methanol, proceeding with inversion of configuration to give the phosphine oxide, **LXXXIII**. This mechanism contrasts with that suggested by Smith and Trippett (141).*

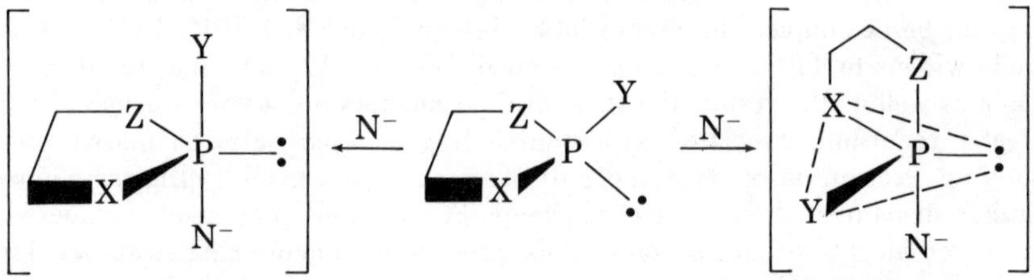

Inversion path   Retention after BR

LXXX   **LXXXI**

P(V)-phosphetanes. Shown in Figure 2.22 is the reaction of an isomer of **LXXXII** with sodium methoxide in methanol, giving the intermediate methyl phosphinite. Rearrangement of the latter phosphinite produces the isomeric phosphine oxide, **LXXXIII** (*141, 142*).

While ring strain may be reduced further in **LXXXIV**, the presence of substituents of contrasting electronegativities is expected to exert an over-riding influence in allowing preferential placement of the more electronegative groups in axial positions. Pseudorotation of **LXXXIV** to bring the chloride

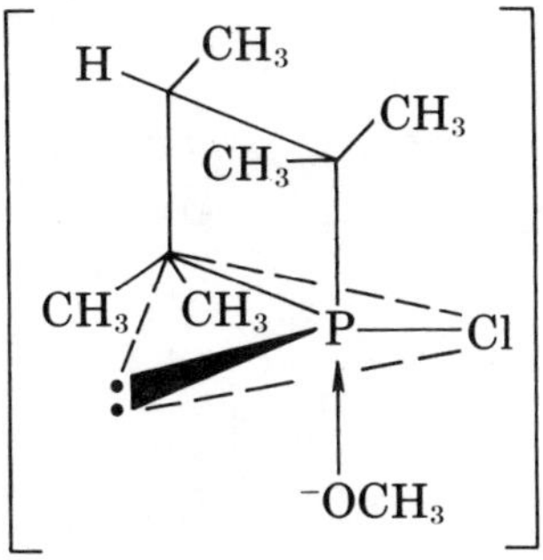

LXXXIV

ion into a leaving position would lead to retention of configuration. However, the observed inversion of configuration (*141*) shows that this mechanism is not followed.

For nucleophilic substitutions of **LXXXII**, Smith and Trippett (*141*) propose Intermediate **LXXXV**, with the unconventional requirement that the entering and leaving groups do so from equatorial positions in order to explain the observed inversion. However, the least electronegative groups

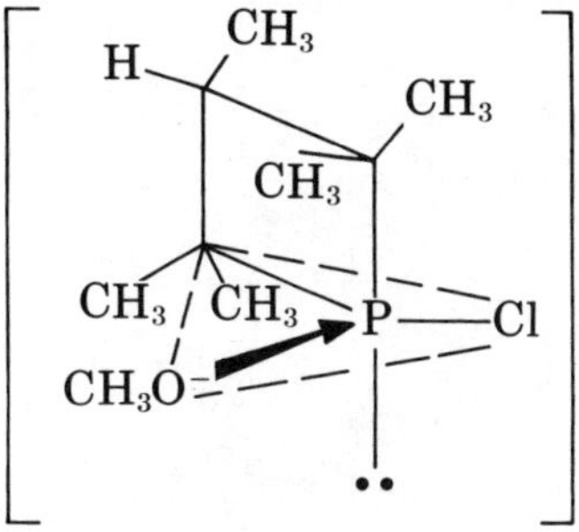

LXXXV

are located in axial positions in this structure. Based on our discussion, this formulation is energetically unfavorable (*143*). Perhaps examination of additional nucleophilic substitutions of **LXXXII** containing less-axiophilic departing groups would reveal crossover to a retention scheme.

Acyclic Tertiary Phosphines. Kyba (*144, 145, 146*) observed that nucleophilic substitution by alkyllithium reagents at chiral phosphorus centers in tertiary phosphines proceeds with inversion of configuration. For the benzyl derivative, **LXXXVI**, nucleophilic substitution by *n*-BuLi and *tert*-BuLi in a variety of solvent media showed only inversion (*144*). The results are consistent with attack opposite the departing benzyl group (Figure 2.23a). Using groups of considerably poorer leaving ability than benzyl (e.g., phenyl and *p-tert*-BuC$_6$H$_4$, which have leaving abilities comparable with each other), it was reasoned (*145*) that the lifetimes of intermediates might be increased

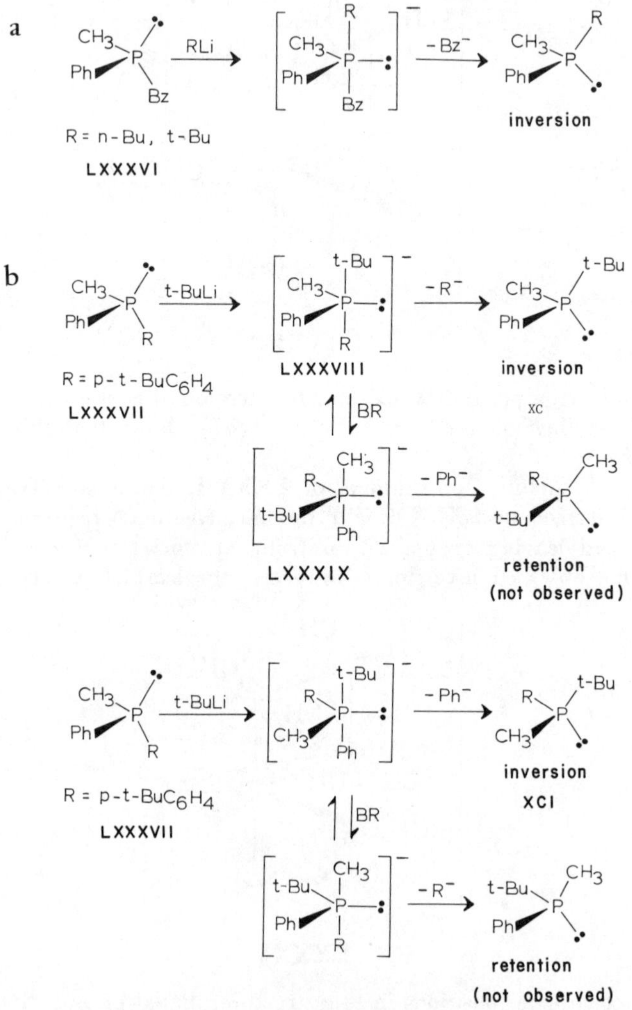

Figure 2.23. *Nucleophilic substitution at phosphorus in tertiary phosphines via trigonal bipyramidal intermediates. (a) Attack by BuLi on (+)-(R)$_p$-benzylmethylphenylphosphine,* **LXXXVI,** *with inversion of configuration (144); (b) attack by tert-BuLi on (+)-(S)$_p$-methylphenyl-p-tert-butylphenyl-phosphine,* **LXXXVII,** *via possible inversion and retention pathways (145). Products* **XC** *and* **XCI** *were observed as expected for the inversion route. Intermediates* **LXXXVIII** *and* **LXXXIX** *may form enantiomeric conformations by a minimum of five pseudorotations and lead to a racemic product, again not observed. In at least one of the pseudorotated intermediates, the lone electron pair, which has the lowest electronegativity, must occupy an unfavorable axial position. Refer to p 30 for a topological description of this epimerization (* **12** *to* **21** *), showing the five required pseudorotations.*

Table 2.12. Rate Constants (L mol^{-1} min^{-1}) for Reactions of XCII (k_c) and $(CH_3O)_2PN(CH_3)_2$ (k_a) in Nitrobenzene at 34°C (*147*)

Reactant	XCII	$(CH_3O)_2PN(CH_3)_2$	k_c/k_a
Methyl iodide	0.03	0.097	0.33
Benzaldehyde	0.11	0.97×10^{-4}	1156
Phenyl isocyanate	0.38	1.12×10^{-3}	339

Chemical Communications

sufficiently to allow equilibration via pseudorotation between them. However, the phosphine products, **XC** and **XCI**, obtained from the reaction of **LXXXVII** with *tert*-BuLi are consistent with the inversion pathway and not with the retention route (*see* Figure 2.23b) (*145*).

More Complex Systems. In the realm of more complex reactions, typified by the interaction of a cyclic phosphoramidite with nucleophiles, Greenhalgh and Hudson (*147*) observed enhanced reactivity compared with the analogous acyclic compound. For example, 2-dimethylamino-1,3,2-dioxaphospholane, **XCII**, reacted ~1100 times faster than dimethyl *NN*-dimethylphosphoramidite, $(CH_3O)_2PN(CH_3)_2$, with benzaldehyde in nitrobenzene solution (Table 2.12). Speculating that a pseudopentacoordinated state arises in the rate-determining step (*38, 147*), both the electronegativity rule and the formation of a spirocyclic structure suggests that relief of ring strain proceeds via **XCIII** as the most stable intermediate formulation (Figure 2.24) (*38, 148*). In keeping with previous mechanisms, pseudorotation allows the leaving group to orient in an axial position. Thus, the postulated mechanism is of the retention type.

In contrast, the cyclic and acyclic phosphoramidite react at comparable rates with methyl iodide (Table 2.12). In this case, the postulated mechanism

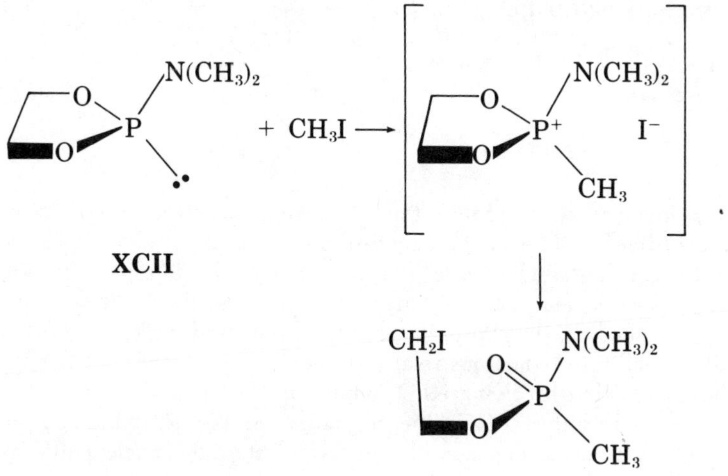

Chemical Communications (*147*)

Table 2.13. Rate Constants for the Michaelis-Arbuzov Reaction of Ethyl Iodide With Cyclic and Acyclic Phosphites in Acetonitrile

Compound	Concentration (mol/L)		Temp (°C)	Second-Order Rate Constant (L mol⁻¹ sec⁻¹ × 10⁵)
	Phosphite	Ethyl iodide		

Compound	Phosphite	Ethyl iodide	Temp (°C)	Rate Constant
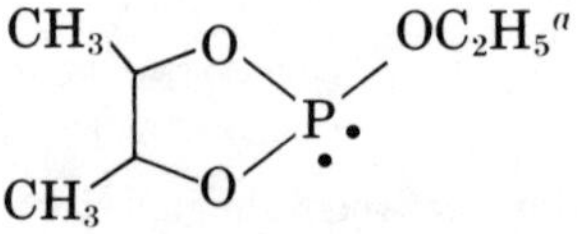	0.296	0.246	70.0	0.7 ± 0.2
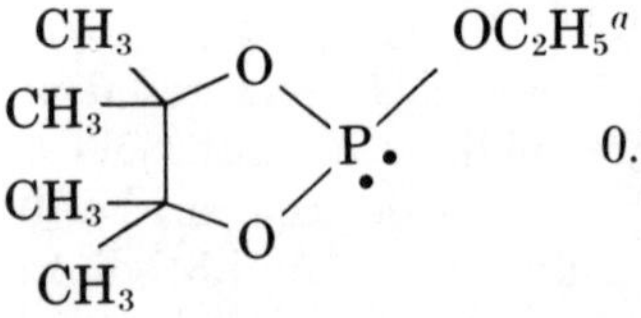	0.121	0.356	70.0	1.1 ± 0.3
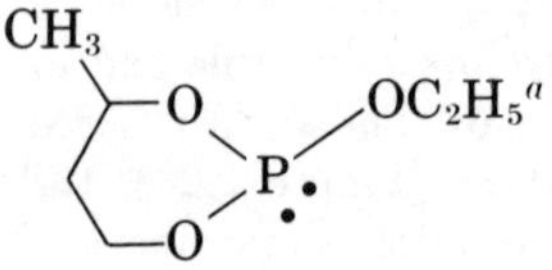	0.187	0.206	70.0	2.7 ± 0.4
	0.229	0.247	65.0	1.5 ± 0.4
	0.273	0.287	60.0	0.9 ± 0.2
Triethylphosphite[b]	0.1996	0.7296	69.9	4.80
Triisopropylphosphite[c]	0.139	0.139	69.9	7.96

[a] Ref *149*.
[b] Ref *150*.
[c] Ref *151*.

(*147*) involves electrophilic attack on the phosphorus atom, leading to a tetracoordinated state in the rate-determining step. Because of a hybridization change, a slight increase in ring strain in the intermediate is expected (*147*), in agreement with the reduced reactivity of the cyclic structure.

In related cyclic phosphite esters, Aksnes and Eriksen (*149*) reported similar small rate reductions relative to acyclic compounds for the Michaelis–Arbuzov reaction involving ethyl iodide in acetonitrile (Table 2.13). Again the rate-determining step is the formation of the phosphonium salt while product distribution depends on the second step, the nucleophilic attack by iodide ion in the phosphonium salt on one of the carbons linked to oxygen. As methyl substitution on the ring is increased, less ring cleavage is observed.

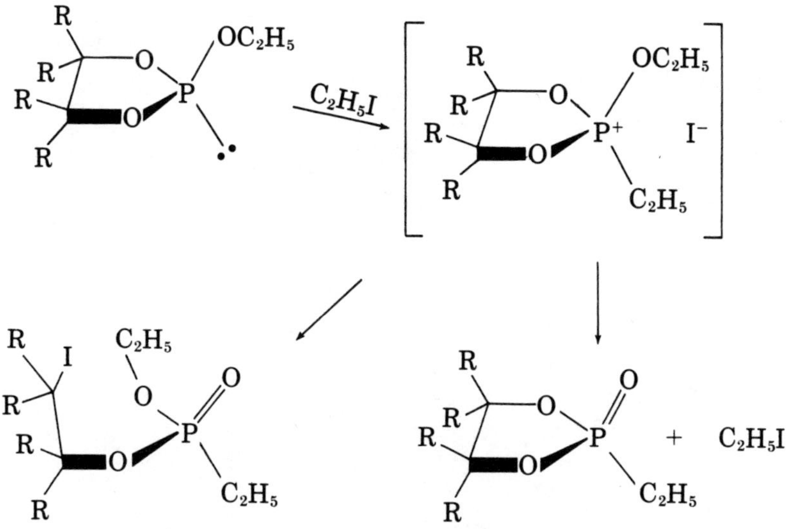

Acta Chemica Scandinavica (*149*)

With reference to the reaction of the cyclic phosphoramidite, **XCII** with benzaldehyde, an alternative mechanism (*152, 153*) has been proposed, involving initial quaternization at phosphorus. However, the rate data comparison between cyclic and acyclic structures appears to favor the argument for the pentacoordinate intermediate shown in Figure 2.24. Also some satisfaction is derived from the fact that stable five-coordinated structures related to the type proposed are known (*154*) and, in a limited number of cases, have had their structural parameters determined by x-ray diffraction (Volume I, p 24) (*126*).

Reactions of isocyanates can be envisioned to follow similar courses for nucleophilic and electrophilic substitutions (*147, 155, 156*) and lead to self-consistent rate interpretations. However, in general, reactions involving compounds containing both nitrogen and oxygen attached to phosphorus are more varied, and a variety of products are obtained with subtle changes in reaction conditions and substituents (*34, 38, 155, 157, 158, 159*). Also, the ready formation of equilibria involving zwitterions lends increased uncertainty in the suggested course of some of these reactions (Volume I, p 185).

Radical Reactions. Reactions of trivalent phosphorus compounds with radicals represent systems conducive to the formation of pseudopentacoordinate intermediates (*160, 161*). Bentrude and co-workers (*162*), for example, studied the reaction of the isomeric cyclic phosphites (2-methoxy-5-*tert*-butyl-1,3,2-dioxaphosphorinanes, **XCIV** and **XCV**) with *tert*-butoxy and *n*-butylthiyl radicals. From a comparison of reactant phosphite and product oxide ratios, it was concluded that the oxidation proceeded stereospecifically with retention of configuration. Similar experiments on the oxidation of optically active methyl-*n*-propylphenylphosphine by *tert*-butoxyl radical likewise showed

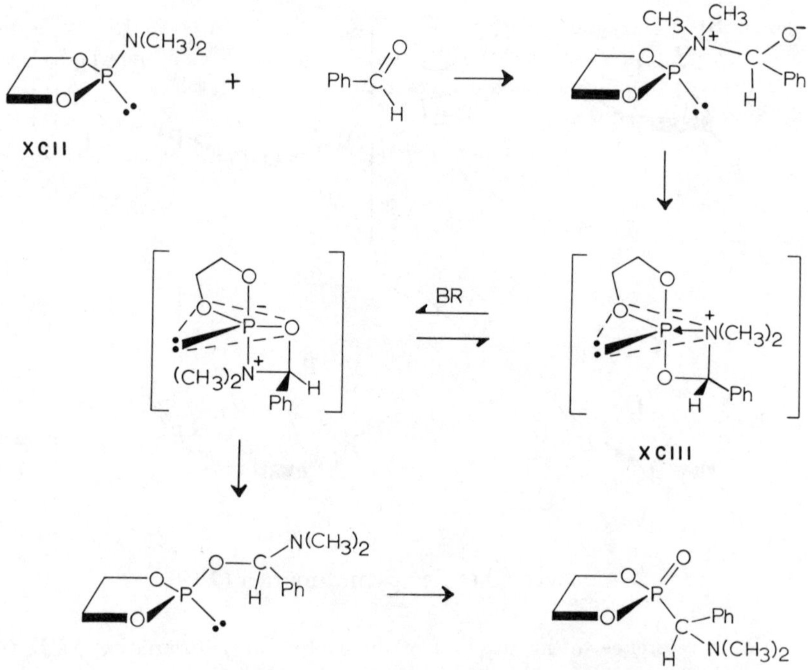

Figure 2.24. A possible mechanism for the reaction of 2-dimethylamino-1,3,2-dioxaphospholane, **XCII**, with benzaldehyde in nitrobenzene solution (38, 147, 148)

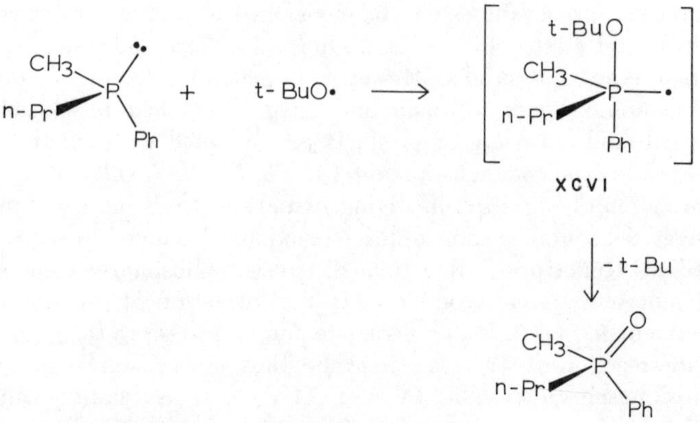

Figure 2.25. Stereospecific oxidation of optically active methyl-n-propyl-phenylphosphine by tert-butoxyl radical proceeding with retention of configuration (162)

a stereospecific reaction with retention of configuration (Figure 2.25) (*162*). Since the entering and departing groups occupy the same axial position in the proposed pseudopentacoordinated intermediate **XCVI**, retention results without an intervening pseudorotation. This contrasts with the general retention mechanism for reactions that do not involve radical attack (Figure 2.1a).

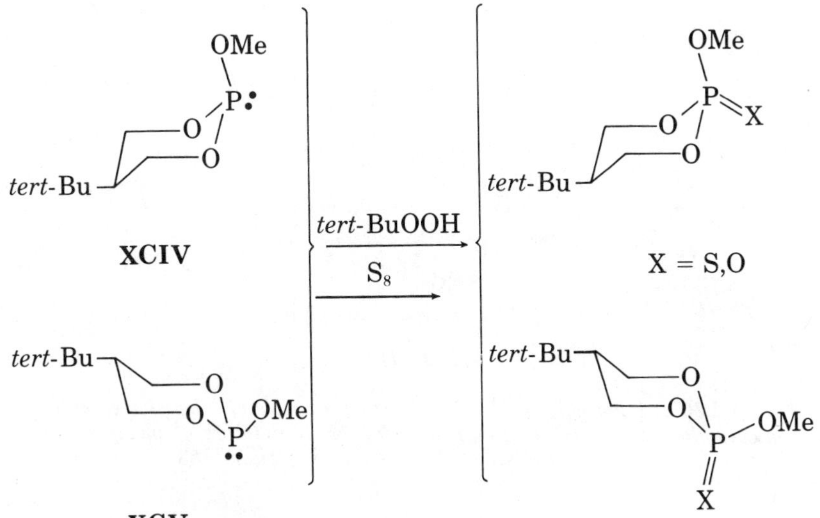

A mechanism like that shown for the phosphine in Figure 2.25 is also possible for the cyclic phosphites.

In another study, on the reaction of ^{14}C-labeled *tert*-butoxy radicals with tri-*tert*-butyl phosphite, the results suggest an irreversible formation of a phosphoranyl radical intermediate (*163*). Because approximately 75% of the product phosphate was ^{14}C-labeled, the intermediate must allow a statistical scrambling of the label. This may be accomplished by pseudorotation of the initial trigonal bipyramidal Intermediate, **XCVII** (odd electron as the pivotal "ligand"), followed by β-scission, as represented in Figure 2.26. Pseudorotation is expected to be a low-energy process for Intermediate **XCVII** compared with that for **XCIV** (Figure 2.25), owing to the presence of four identical groups. It is recalled (Volume I, p 54) that low-temperature ESR data on phosphoranyl radicals containing alkoxy groups support a trigonal bipyramidal structure that undergoes rapid equilibration of the alkoxy groups as the temperature is increased. If we assign a low electronegativity value to the odd electron (about the same as that for an alkyl group (Table 2.1)), 1.4 kcal/mol is estimated for the barrier in the pseudorotational process: **XCVII** ⇌ **XCVIII**.

Because reactions involving phosphoranyl intermediates are not as well studied as those involving phosphoranes (Volume I, p 51), greater caution must be exercised in their mechanistic interpretations. For example, instead of the structure shown for **XCVII**, although it is the preferred one, a tetrahedral or square pyramidal intermediate will allow configurational equivalency of

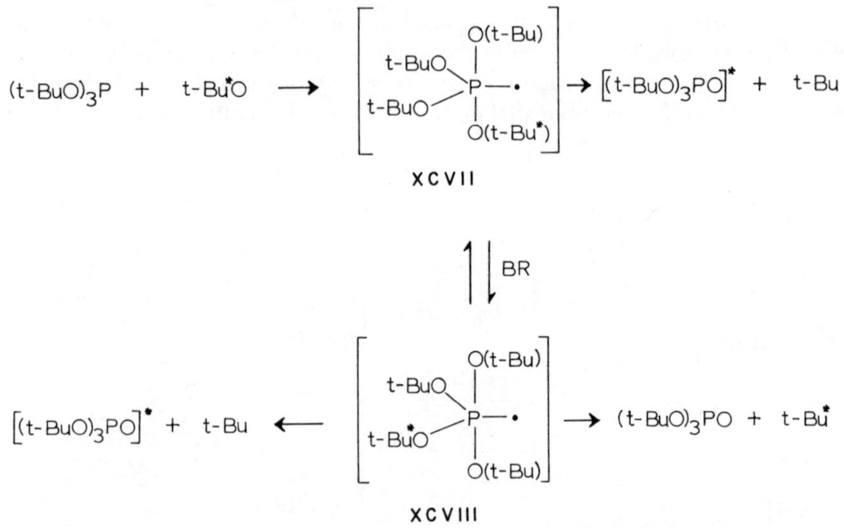

$$(t\text{-}BuO)_3P \ + \ t\text{-}Bu\overset{*}{O} \ \longrightarrow \ \begin{bmatrix} & & O(t\text{-}Bu) \\ t\text{-}BuO & & | \\ & \diagdown \!\!\!\diagup P\!\!-\!\!\bullet \\ t\text{-}BuO & & | \\ & & O(t\text{-}Bu^{*}) \end{bmatrix} \longrightarrow \ \big[(t\text{-}BuO)_3PO\big]^{*} \ + \ t\text{-}Bu$$

XCVII

⇅ BR

$$\big[(t\text{-}BuO)_3PO\big]^{*} \ + \ t\text{-}Bu \ \longleftarrow \ \begin{bmatrix} & & O(t\text{-}Bu) \\ t\text{-}BuO & & | \\ & \diagdown \!\!\!\diagup P\!\!-\!\!\bullet \\ t\text{-}B\overset{*}{u}O & & | \\ & & O(t\text{-}Bu) \end{bmatrix} \longrightarrow \ (t\text{-}BuO)_3PO \ + \ t\text{-}B\overset{*}{u}$$

XCVIII

*Figure 2.26. A possible mechanism for the reaction of labeled tert-butoxy radical wih tri-tert-butylphosphite involving pseudorotation of phosphoranyl Intermediate **XCVII**, followed by β-scission (163)*

the *tert*-butoxy groups to be achieved (*163*). A less likely possibility is that the substituents in **XCVII** might be reactionally equivalent.

As well as greater uncertainty in structuring intermediates, many phosphoranyl radicals formed from compounds of tervalent phosphorus decompose by competitive α- and β-scission (*160, 161, 164–169*). The radical reactions

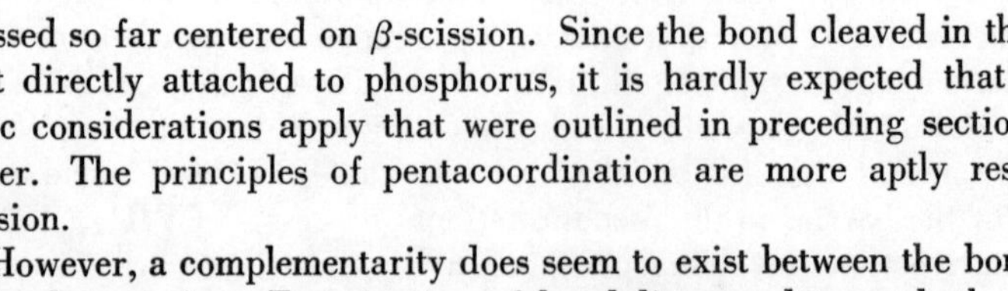

$$RO\bullet + PX_3 \rightarrow RO\overset{\bullet}{P}X_3 \begin{cases} \xrightarrow{\ \alpha \text{ scission}\ } ROPX_2 + X\bullet \\ \\ \xrightarrow[\ \beta \text{ scission}\ ]{} X_3PO + R\bullet \end{cases}$$

discussed so far centered on β-scission. Since the bond cleaved in this process is not directly attached to phosphorus, it is hardly expected that the same kinetic considerations apply that were outlined in preceding sections of this chapter. The principles of pentacoordination are more aptly reserved for α-scission.

However, a complementarity does seem to exist between the bond cleaved in α- and β-scission. Examination of bond-distance data on both cyclic and acyclic phosphoranes (Volume I, p 12ff) suggests that although the axial P–X bond is weaker than a similar equatorial P–X bond in a trigonal bipyramid, the adjacent bonds bear an opposite bond-strength relationship—i.e., the adjacent bond (X–Y) in the axial sequence P–X–Y is stronger than the related equatorial linkage (6, 7, 126). According to this view, α-scission is expected to occur preferentially from an axial position of a trigonal-bipyramidal intermediate, and β-scission is expected to occur preferentially from an equatorially attached substituent. Later, we examine experimental data that led to a similar proposal (166, 167).

There are reasons to believe that the scission processes should not take place exclusively in the manner just cited. First β-scission involves ligand bonding considerations that result as a secondary effect of bonding differences between the two positions of the trigonal bipyramid (126). Secondly, the structures of phosphoranyl intermediates seem to undergo greater variations with ligand substitution than the structures of phosphorane derivatives. For example, ESR evidence indicates that the phosphoranyl radical, $PhP(OMe)_2$ (O-*tert*-Bu), has a tetrahedral configuration, largely because of the delocalization of the odd electron into the phenyl ring (170). This contrasts with tetra-alkoxyphosphoranyl radicals whose structures are indicated to be nearly trigonal bipyramidal with some degree of distortion toward a square pyramid (171).

Therefore, the positional preferences discussed for α- and β-scission, although they may be followed in general, probably involve both sites of the trigonal bipyramid to some degree; the degree depends on the particular make-up of the phosphoranyl radical. However, as the structures of the radicals tend toward tetrahedral and less discrimination between sites becomes possible, the importance of the point of attack on the phosphoranyl radical as a mechanistic determinant should take over. Here the steric and electronic properties of the various substituents relative to that of the attacking radical should be of utmost significance.

In this regard, the intermediate in Figure 2.25 probably has a structure distorted toward a tetrahedral configuration, in view of ESR data supporting the location of the lone electron on the phenyl ring in the related radical, $PhP(OMe)_2$(O-*tert*-Bu) (170). Attack by the *tert*-butoxy group at a tetrahedral face opposite the phenyl group is most reasonable in terms of both steric and electronic effects. For the intermediate in Figure 2.25, the pseudorotational barrier should be relatively high since alkyl groups will be placed in axial positions (11.8 kcal/mol is estimated by our model, p 88, for the process with the lone electron as the pivotal ligand). Hence, by either mechanism, retention of configuration in forming the phosphine oxide represents the preferred course. (Retention observed for radical reactions of the cyclic Derivatives **XCIV** and **XCV** (p 169) implies lack of pseudorotation (166) in a trigonal-bipyramidal intermediate. The six-membered ring is presumed to be oriented diequatorially by analogy with the intermediate structures inferred for Entries **10–13** of Table 2.11. For this ring orientation a low-energy pseudorotation is not possible.)

Application of the hypothesis above concerning positional preferences in α- and β-scission rationalizes the dominance of α-scission in certain reactions. Radical decomposition via β-scission is normally thermodynamically favored because of the large increase in enthalpy that occurs when the phosphoryl bond is formed ($\sim$ 144 kcal/mol (172)). In the reaction of tri-n-butylphosphine with *tert*-butyl peroxide, however, α-scission results in an 80% yield of *tert*-butyl di-n-butylphosphonite compared with the 20% yield of tri-n-butyl-

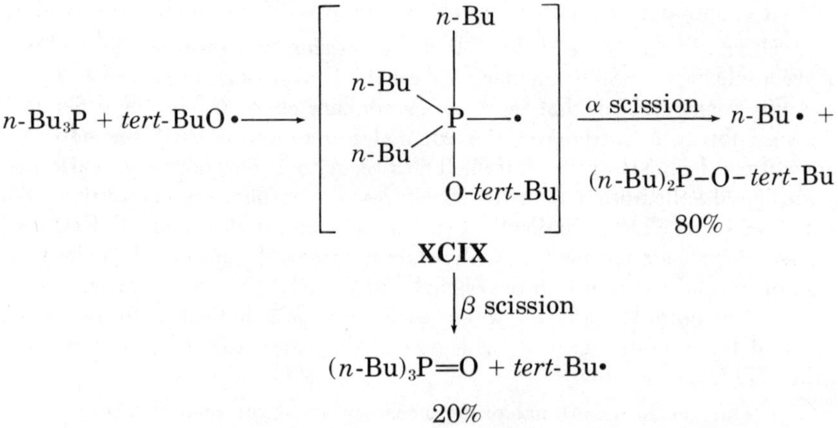

$$n\text{-Bu}_3\text{P} + tert\text{-BuO} \cdot \longrightarrow \left[\begin{array}{c} n\text{-Bu} \\ n\text{-Bu} \diagdown \\ n\text{-Bu} \diagup \text{P} \longrightarrow \cdot \\ \text{O-}tert\text{-Bu} \end{array} \right] \xrightarrow{\alpha \text{ scission}} \begin{array}{c} n\text{-Bu} \cdot + \\ (n\text{-Bu})_2\text{P}-\text{O-}tert\text{-Bu} \\ 80\% \end{array}$$

XCIX

$\Big\downarrow \beta$ scission

$(n\text{-Bu})_3\text{P}{=}\text{O} + tert\text{-Bu}\cdot$

20%

phosphine oxide produced by β-scission (164). For β-scission to occur preferentially at an equatorial site, pseudorotation of **XCIX** is required.

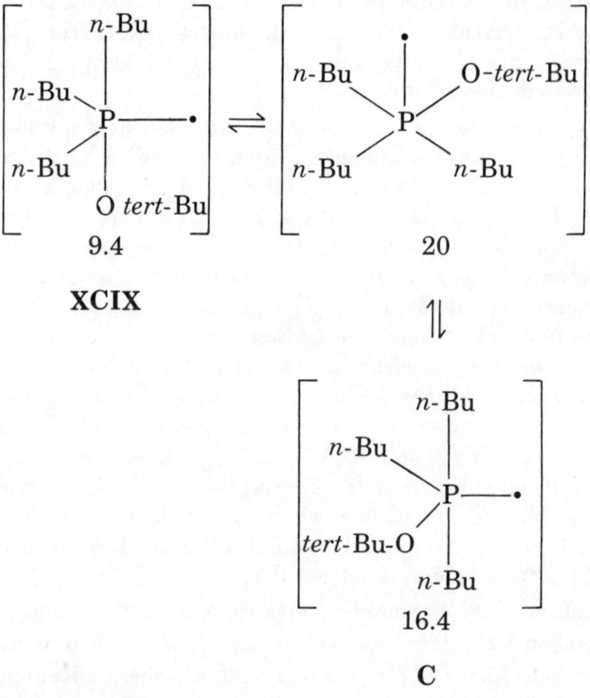

$$\left[\begin{array}{c} n\text{-Bu} \\ n\text{-Bu} \diagdown \\ n\text{-Bu} \diagup \text{P} \longrightarrow \cdot \\ \text{O } tert\text{-Bu} \end{array} \right] \rightleftharpoons \left[\begin{array}{c} \cdot \\ n\text{-Bu} \diagdown \quad \diagup \text{O-}tert\text{-Bu} \\ \text{P} \\ n\text{-Bu} \diagup \quad \diagdown n\text{-Bu} \end{array} \right]$$

9.4   20

XCIX $\Updownarrow$

$$\left[\begin{array}{c} n\text{-Bu} \\ n\text{-Bu} \diagdown \\ tert\text{-Bu-O} \diagup \text{P} \longrightarrow \cdot \\ n\text{-Bu} \end{array} \right]$$

16.4

C

This should be a relatively high-energy process since **C** has the *tert*-butoxy group and an additional *n*-butyl group interchanged from their preferred orientation in **XCIX**. We estimate 16.0 kcal/mol for this barrier (*see* p 89) compared with 1.4 kcal/mol estimated for pseudorotation of **XCVII**, the latter containing four *tert*-butoxy groups (Figure 2.26).

Support for the operation of α- and β-scission by way of axial and equatorial sites, respectively, comes from ESR studies by Davies and co-workers (*166, 173*). The high stability (*173*) toward unimolecular scission of the radical $Et_2\dot{P}(O\text{-}tert\text{-}Bu)_2$, **CI**, compared with $Et\dot{P}(OEt)_2O\text{-}tert\text{-}Bu$, **CII**, and $Et_3\dot{P}O\text{-}tert\text{-}Bu$, **CIII**, suggested that the ethyl groups and *tert*-butoxy groups in **CI** were in the wrong positions for facile α- and β-scission (*166, 173*). The expected phosphoranyl structures for these radicals are shown below, and the observed order for loss of ethyl radicals by α-scission is

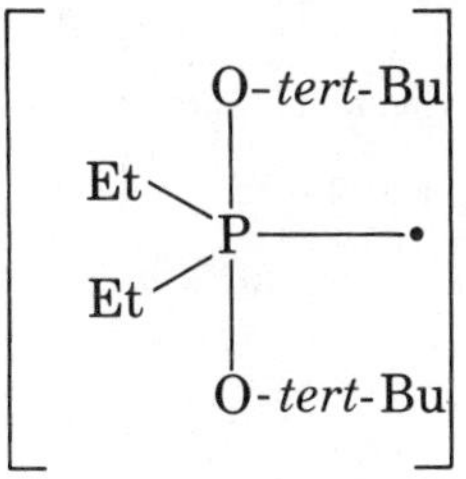

CIII > CII > CI. Pseudorotation of **CI** and **CII** would allow ethyl groups to reach departing axial locations. We estimate the associated barriers as 6.8 kcal/mol for **CII** and 12.2 kcal/mol for **CI**. In agreement with the relatively high pseudorotational barrier expected for **CI**, the lack of β-scission for this radical lends support to the proposal that β-scission occurs preferentially from an equatorial position (*166, 167*). The diphenoxy radical, $(CH_3)_2\dot{P}(OPh)_2$, also has been reported to have an unusually high stability (*174*). An ESR study of the kinetics of α-scission of alkylalkoxyphosphoranyl radicals conducted at low temperatures, where β-scission is noncompetitive, resulted in rate constants for the decay of Isomer **CIV** (*175*). These are in Table 2.14. If α-scission did occur by loss of the alkyl group from an equa-

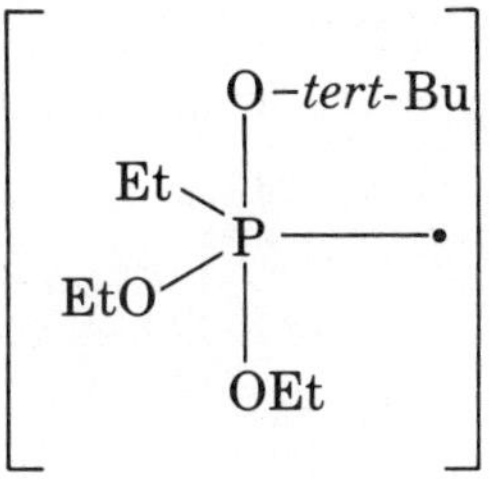

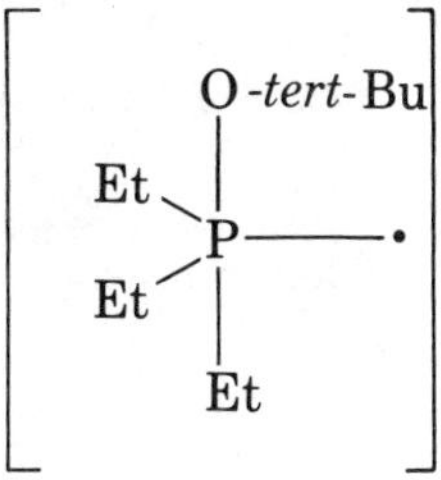

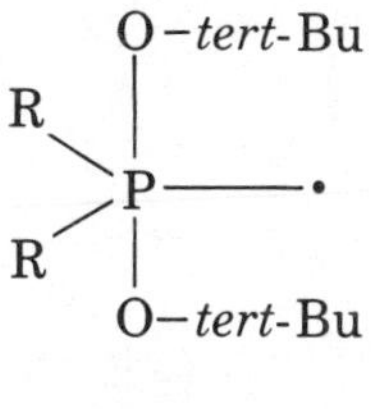

CIV

Table 2.14. Kinetics of α-Scission of Alkylalkoxyphosphoranyl

Phosphoranyl Radical[a]	Solvent	Temp Range (°C)
$Me_2\dot{P}(OBu^{tert})_2$	Isopentane[b]	−55 to −82
$Et_2\dot{P}(OBu^{tert})_2$	Isopentane[b]	−76 to −102
$Pr^n_2\dot{P}(OBu^{tert})_2$	Propane	−73 to −109
$Pr^i_2\dot{P}(OBu^{tert})_2$	Propane	−95 to −120
$Bu^s_2\dot{P}(OBu^{tert})_2$[c]	Propane	−95 to −120
$Bu^{tert}_2\dot{P}(OBu^{tert})_2$	Propane	−70 to −101
$(Allyl)_2\dot{P}(OBu^{tert})_2$	Propane	—
$Me_2\dot{P}(OEt)OBu^{tert}$	Isopentane	−48 to −93
$Et_2\dot{P}(OEt)OBu^{tert}$	Isopentane	−96 to −128
$Et_2\dot{P}(OEt)OBu^{tert}$[e]	Isopentane	−37 to −88
$Et\dot{P}(OEt)_2OBu^{tert}$	Propane	−150 to −170

[a] Rate of α-scission determined by direct decay of ESR signal of phosphoranyl radical unless otherwise stated.
[b] Rate in propane at −100°C was the same within experimental error.
[c] The two diastereoisomeric forms appear to decay at similar rates.
[d] Rate constant at −140°C. Decay at −100°C was too rapid to follow.

torial site, a rate-order increase consistent with a decrease in the P–C bond strength might be expected—i.e., the rate order should be R = Me < Et ~ n-Pr < iso-Pr ~ sec-Bu < tert-Bu < allyl. However, the observed order places tert-$Bu_2\dot{P}$(O-tert-Bu)$_2$ intermediate in reactivity between $Me_2\dot{P}$(O-tert-Bu)$_2$ and $Et_2\dot{P}$(O-tert-Bu)$_2$. Pseudorotational processes bringing the alkyl groups to departing axial positions, with steric hindrance implied for this process when dealing with the bulky tert-butyl groups (175), makes the observed rate order more comprehensible.

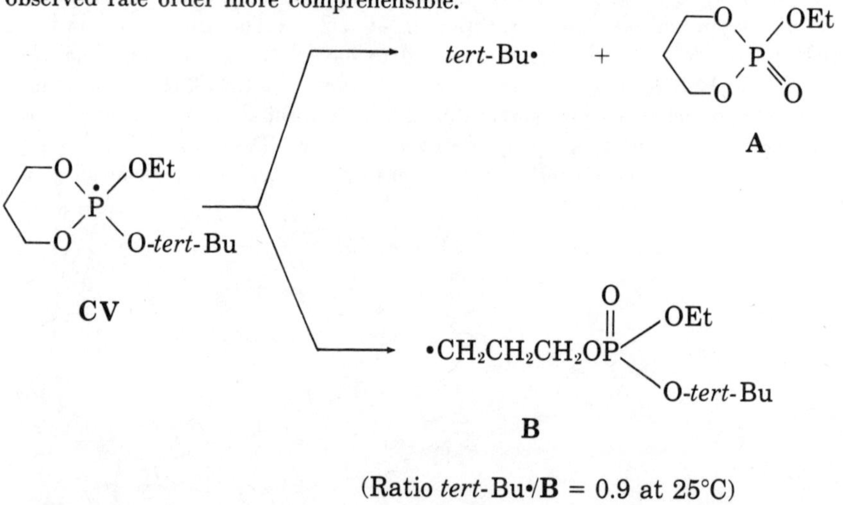

(Ratio tert-Bu•/B = 0.9 at 25°C)

Journal of the Chemical Society (166)

Radicals in Solution (175)

A/sec^{-1}	$E_a,$ $kcal/mol$	Rate Constant for α-Scisson at $-100°C$ (sec^{-1})
4.0×10^{13}	14.0	9.5×10^{-5}
1.5×10^{11}	10.5	8.3×10^{-3}
7.5×10^{10}	10.4	6.5×10^{-3}
2.9×10^{12}	10.9	6.0×10^{-2}
4.6×10^{11}	10.1	7.0×10^{-2}
2.0×10^{12}	12.0	1.4×10^{-3}
—	—	0.35^d
9.0×10^{12}	13.6	6.0×10^{-5}
4.0×10^{10}	8.2	2.0
1.1×10^{10f}	7.8^f	1.6^f
2.2×10^{8g}	4.4^g	630^h

^{e} Rate constant measured by the "steady state" method (*175*).
^{f} Taking $2k_{10} = 2.6 \times 10^{10} \exp(-0.83/RT)$ L mol^{-1} sec^{-1}, where RT is in kcal/mol.
^{g} Probably very inaccurate due to small temperature range available within which decay was first order and low signal strength.
^{h} Rate constant at $-160°C$ is 0.55 sec^{-1}.

Journal of the Chemical Society

The six-membered dioxa ring-containing phosphoranyl radical, **CV**, undergoes competing β-scission with partial ring opening, whereas no ring cleavage was detected from β-scission of the corresponding five-membered dioxa radical, **CVI** (*166*). The placement of the five-membered ring in an axial–equatorial configuration and the six-membered ring in a diequatorial location is consistent with structural principles of phosphorane intermediates (pp 149, 155) and appears to apply here as well. Rapid pseudorotation of **CVI**

CVI   CV

(lone electron as the pivot) allows loss of *tert*-Bu · to occur preferentially at an equatorial site. The reported (*166*) rate constant (at $-60°C$) for β-scission of **CVI**, $k = 43.6$ sec^{-1} in cyclopentane, is less than that for the acyclic analog $(EtO)_3PO\text{-}tert\text{-}Bu$, $k = 1.8 \times 10^2$ sec^{-1} at $-60°C$ in cyclopentane, suggesting some degree of ring strain in Phosphoranyl Radical **CVI** since the pseudorotational processes should be comparable in the two cases (*166*). In the cyclic radical, **CVII**, analogous to **CV**, only ring cleavage was detected above $-50°C$

CVII

(166). The relatively high pseudorotational barriers expected for Structures **CV** and **CVII** support the occurrence of ring opening at an equatorial site.

In a later study, on the oxidation of substituted five- and six-membered, cyclic trivalent compounds, **CVIII–CXI**, Bentrude and co-workers (176) generated chiral phosphoranyl radicals. A study of the mode of decomposition of the phosphoranyl radicals produced by *tert*-butoxy radical oxidations of Dioxaphospholane **CVIII** and Dioxaphosphorinane **CX** revealed a nearly stereospecific course with retention of configuration in the product methyl phosphates. Ethoxy radicals react with the *tert*-butoxy Derivatives, **CIX** and

R = CH$_3$O   **CVIII**   R = CH$_3$O   **CX**

R = *tert*-BuO   **CIX**   R = t-BuO   **CXI**

CXI, again nearly stereospecifically but with inversion of configuration at phosphorus in the product ethyl phosphates.

These results support a mechanism involving attack by the alkoxy radicals on Phosphine Derivatives **CVIII–CXI**, leading to trigonal-bipyramidal intermediates that do not undergo permutational isomerization (176). A pseudorotational barrier of at least 11 kcal/mol is estimated for these intermediates (176). The process is illustrated for Phospholane Derivatives **CVIII** and **CIX** (Figure 2.27) although it is proposed to apply to the Phosphorinanes **CX** and **CXI** as well (176). The thiyl Radical, n-BuS·, follows the same course shown in Figure 2.27a with **CX**. In phosphate formation with retention of configuration (Figure 2.27a), β-scission occurs at an axial site whereas in the radical decomposition yielding phosphates with inverted phosphorus (Figure 2.27b), β-scission results in loss of the *tert*-butyl radical from an equatorial site. The occurrence of β-scission at both axial and equatorial sites received confirmation from the observance of cis- and trans-phosphate products from the reaction of **CIX** and **CXI** with *tert*-butoxy radicals. Further, use of ^{14}C-labeled *tert*-butoxy radicals in the presence of *cis*-**CXI**

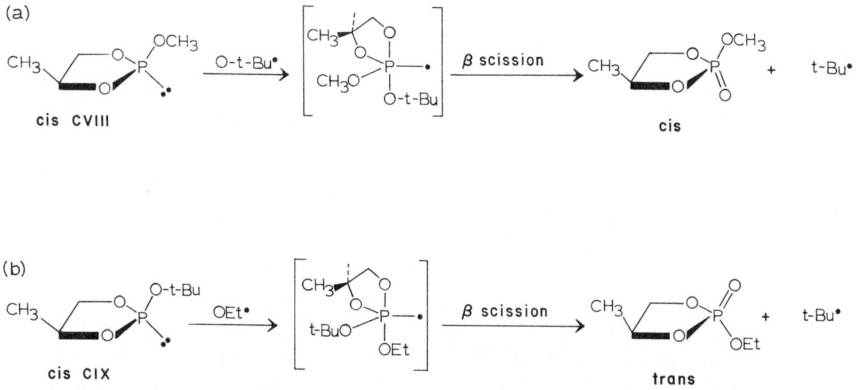

(a)

(b)

Figure 2.27. A mechanism for the reaction of alkoxy radicals with the 2-alkoxy-4-methyl-1,3,2-dioxaphospholanes, **CVIII** and **CIX**. The cis isomer is used for illustration. (a) β-scission with retention of configuration occurs, (b) inversion of configuration at chiral phosphorus is obtained. The same mechanism may be applied as well to the radical oxidations with the 2-alkoxy-5-tert-butyl-1,3,2-dioxaphosphorinanes **CX** and **CXI** (176).

gave both labeled and unlabeled products (*176*). In agreement with the mechanism in Figure 2.27b, only the trans isomer was labeled, indicating the absence of pseudorotation.

The foregoing results on the decomposition of the phosphoranyl radicals formed from alkoxy and thiyl radical oxidations of Phosphines **CVIII–CXI** are also consistent with a mechanism involving diequatorial ring placement in the phosphoranyl intermediate where oxide formation takes place with lack of pseudorotational processes (Figure 2.28). If we continue our comparison between phosphoranyl radicals and phosphorane intermediates, an analogy is apparent (pp 149ff, 155), which lends support to the mechanism in Figure 2.27 for radical oxidation of the five-membered cyclic derivatives, **CVIII** and **CIX**, and that given in Figure 2.28 which would apply to the radical oxidation of the six-membered cyclic derivatives, **CX** and **CXI**. (ESR data supported an axial–equatorial orientation in the six-membered cyclic diazaphosphoranyl radicals (*177*).

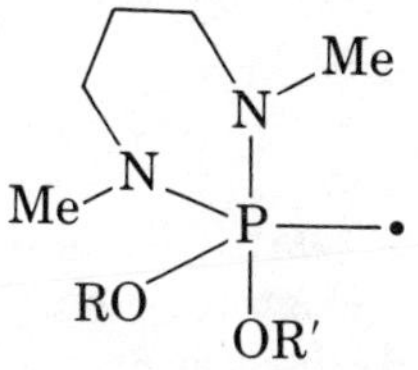

$$R = R' = Et$$
$$R = tert\text{-}Bu, R' = Et$$
$$R = R' = tert\text{-}Bu$$

However, this orientation is favored in this case since the equatorial nitrogen can achieve maximum π bond stabilization (p 29).

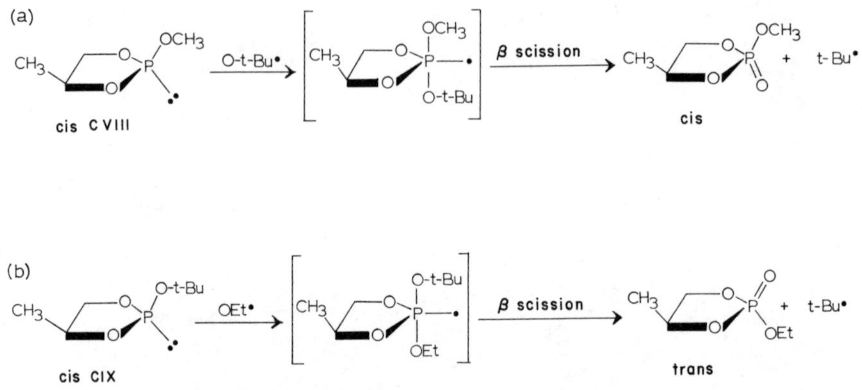

Figure 2.28. A mechanism for the reaction of alkoxy radicals with the 2-alkoxy-4-methyl-1,3,2-dioxaphospholanes, **CVIII** and **CIX** (176). The cis isomer is used for illustration. (a) β-scission with retention of configuration occurs; (b) inversion of configuration at chiral phosphorus is obtained. The same mechanism may be applied as well to the radical oxidations with the 2-alkoxy-5-tert-butyl-1,3,2-dioxaphosphorinanes, **CX** and **CXI**.

Pseudorotation of phosphoranyl intermediates derived from Phospholanes **CVIII** and **CIX** might be expected for the mechanism favored in these cases (Figure 2.27). However, the presence of bulky *tert*-butoxy groups (cf. pp 176, 119) and ring substitution (p 153) are factors that restrict the pseudorotational process. Furthermore, any changes in ring puckering that accompany pseudorotation for the saturated five-membered rings can contribute to the barrier energy.

As alluded to earlier, studies on phosphoranyl radicals are yet in an early stage compared with the systematics that have evolved from the much wider literature on phosphoranes. We have outlined some of the major effort with these interesting radicals and have attempted to correlate the fields in what seems, at present, a reasonable approach. It remains to establish the significance of this comparison. Additional investigations concerned with further elaboration of the basic scission processes, their stereochemistry, and use of ligands of varying axiophilicity should be consulted for a greater appreciation for the subject (*178–183*).

Enzyme Systems

Ribonuclease Catalysis. MODEL SYSTEM. In several well-studied enzyme systems concerned with phosphoryl group transfer, mechanisms have been developed in which pentacoordinated phosphorus is proposed as the transition state species. The presence of enzyme constraints at the active site may be expected to facilitate the attainment of the required transition state (*184*) to allow for their known catalytic behavior. One such system that has

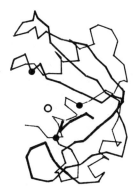

Figure 2.29. Schematic of the structure of ribonuclease A. The location of the active site crevice is indicated by the open circle. Approximate positions of His 12, His 119, and Lys 41 are shown by filled circles.

received much attention (*185*) concerns the action of pancreatic ribonuclease (RNase) on the decomposition of ribonucleic acids.

Ribonuclease A (Figure 2.29) has an active-site crevice across which a ribonucleic acid chain (Figure 2.30) fits (*186*). Subsequent enzymatic action by RNase causes the RNA chain to be cut in a specific manner, as indicated in Figure 2.30. Interactions prominent at the active site involve the amino-acid residues, lysine 41, histidine 12, and histidine 119. Chemical and physical evidence indicating the importance of these residues was strengthened by crystal structure determinations of RNase A (*186, 187, 188, 189*) and RNase S (*186, 190, 191, 192, 193*). A way in which these residues aid in the catalytic action is embodied in a mechanism proposed by Roberts et al. (*194*) (Figure 2.31), based on pentacoordinate principles. Consistent with available data (*185*), a two-part mechanism is described for ribonuclease catalysis (*194, 195, 196, 197*): a transesterification leading to a 2′,3′-cyclic phosphate and a subsequent hydrolysis step resulting in cleavage of the cyclic phosphate.

Most mechanisms that have been suggested can be classified according to the geometry of the displacement reaction in the first part of enzymatic

U = uracil

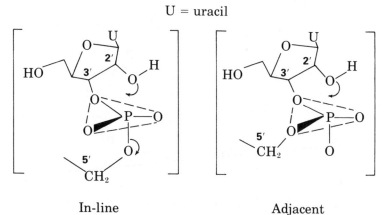

In-line   Adjacent

Proceedings of the National Academy of Sciences U.S.A. (*195*)

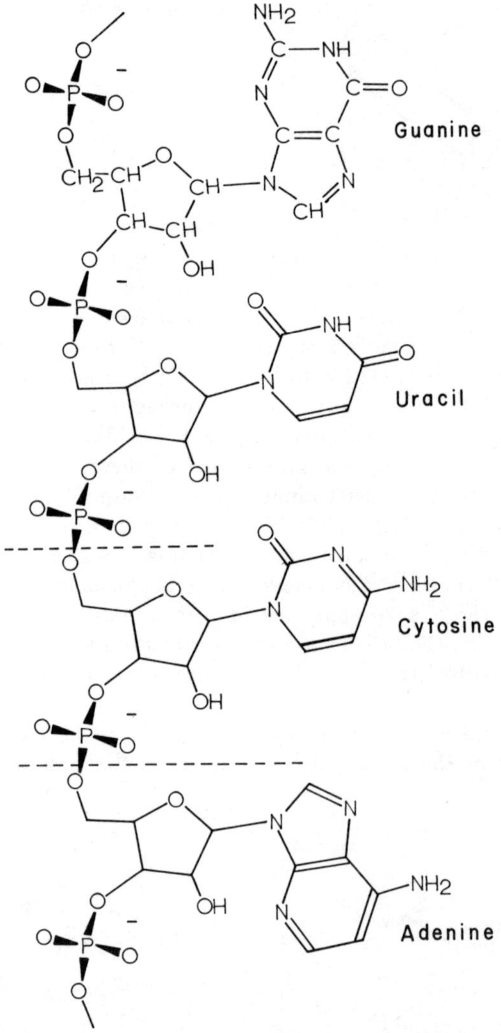

Figure 2.30. Ribonucleic acid chain (RNA). RNase cleaves the P–O linkages as indicated by the dashed lines.

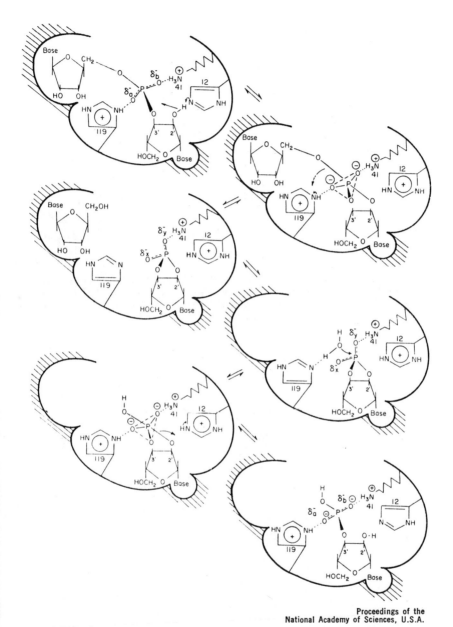

Proceedings of the
National Academy of Sciences, U.S.A.

Figure 2.31. Proposed mechanism of action of RNase (194). The relative positions of the substrate and the amino acid side chains deduced from X-ray crystallography and NMR spectroscopy are shown. Shaded areas represent binding sites for the two nucleoside bases. In the last structure, the ionization states shown are those that result from the reaction; optimal binding of the product results when the phosphate group is doubly ionized and histidine 12 is protonated.

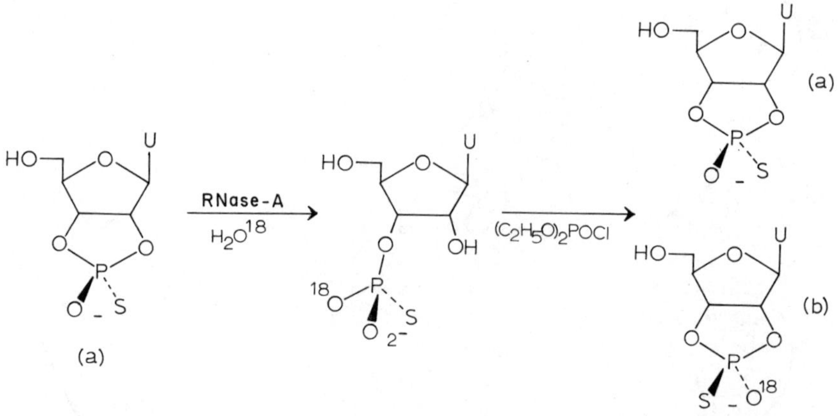

Figure 2.32. An in-line mechanism (198) showing the hydrolysis of chiral uridine-2',3'-cyclic phosphorothioate by RNase using labeled H_2O followed by recyclization with diethylphosphorochloridate. ^{18}O concentrates in the exo Isomer **b**.

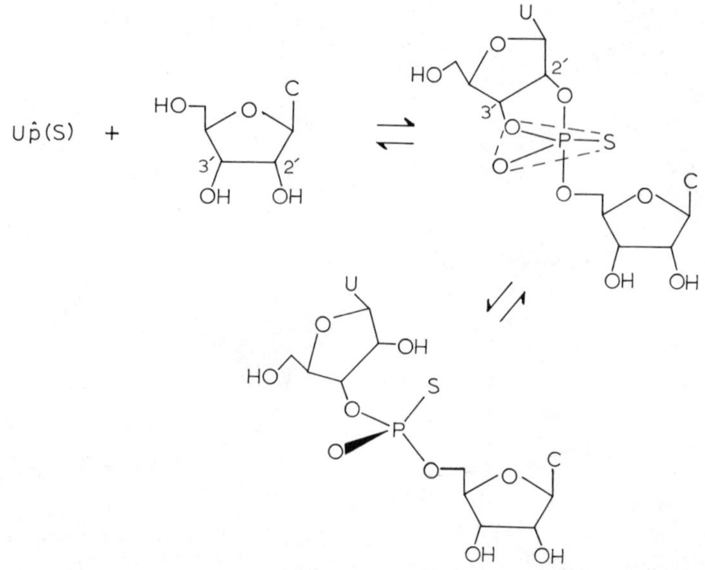

Figure 2.33. RNase A catalyzed synthesis of the dinucleoside phosphorothioate Up(S)C from the crystalline isomer of uridine 2',3'-cyclic phosphorothioate [Up̂(S)] and cytidine by an in-line process (202). This reaction serves as a model for the first part of ribonuclease action.

action. These mechanisms can, in principle, be broken down into two classes (*195*): those in which the incoming and outgoing groups are at axial positions of a trigonal bipyramid (in-line) and those (adjacent) in which it is assumed that the same group acts, first as a general base toward the attacking nucleophile (2′-OH) and then as a general acid toward the leaving group (5′-OCH$_2$). To agree with the preference rules, pseudorotation must occur (*195*) in the adjacent process to bring the 5′-OCH$_2$ group to a leaving axial position. The second part (hydrolysis of the cyclic phosphate) would be similar to a microscopic reverse of the first part, with the leaving 5′-OCH$_2$ group replaced by an entering water molecule. The mechanism (*194*) summarized in Figure 2.31 follows the in-line process.

LABELING EXPERIMENTS. Data pertinent to the second part of ribonuclease catalysis was obtained by Usher and co-workers (*198*), who performed labeling experiments on the model chiral system: uridine-2′, 3′-cyclic phosphorothioate, [Up(S)], **CXII**. Isomer **a** was hydrolyzed (*198*) at pH 7 by pancreatic ribonuclease A in water enriched in ^{18}O (Step 1), and the resulting monoester was recyclized by treatment with diethylphosphorochloridate in

CXII

pyridine solution to give a mixture of Isomers **a** and **b** (Step 2). Subsequent x-ray structural analysis was performed on the triethylammonium salt (*199*). An in-line mechanism representing this process is shown in Figure 2.32. It is assumed (*198, 200*) that enzymic hydrolysis gives only the 3′-monoester as found (*201*) for the related phosphate.

With these basic conditions, the ring closure caused by diethylphosphorochloridate would be expected to proceed by an in-line process (*198*). On the basis of the preference rules, the relatively large differences in electronegativities of the substituents in this medium would require the more electropositive anionic groups to remain in equatorial positions. As a result, pseudorotation would be inhibited. Under this postulation, the enzymic ring opening (Step 1), if it also proceeds in-line would produce Isomer **a** with no excess ^{18}O while Isomer **b** should contain one equivalent of ^{18}O from the enriched water (*198*). On the other hand, if an adjacent mechanism were followed in the enzymic step, the presence of ^{18}O in the product isomers would be reversed. Isotopic analysis (*198*) was essentially in accord with the in-line mechanism showing that for this compound the enzymic hydrolysis and ring closure proceed with the same type of geometry. Accordingly, it was concluded (*198*) that the second part in the action of ribonuclease A is an in-line process.

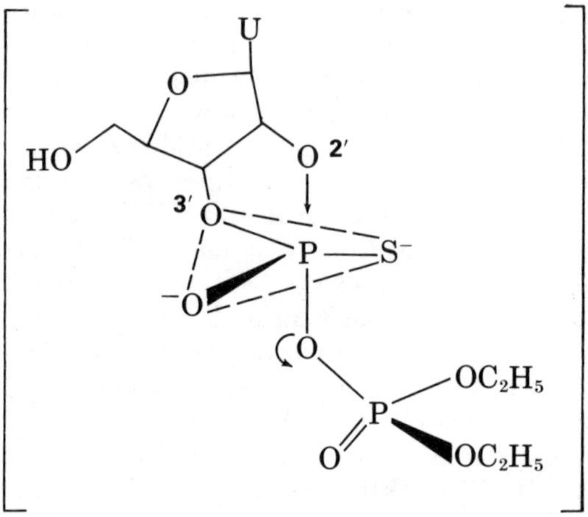

In continuance of their elegant work, Usher et al. (*202*) studied the enzyme-catalyzed synthesis of uridylylthio 3′,5′-phosphorylcytidine Up(S)C from the crystalline isomer [Up̂(S)] and cytidine (Figure 2.33). This synthesis was designed to model the first part of ribonuclease action. The latter is known to proceed reversibly. The cyclic isomer, Up̂(S), was reformed by a nonenzymatic reaction known to proceed in-line. The fact that the original cyclic isomer was obtained and not the one of opposite chirality established an in-line mechanism. By invoking the principle of microscopic reversibility, the first part of RNase action (the transesterification to give cyclic phosphate) is reasoned to proceed in-line (*202*). By inference (*202*), a normal dinucleoside phosphate is expected to behave similarly to the thio derivative and to follow the same stereochemical course. NMR and x-ray evidence are consistent (*194*) with the latter interpretation.

The stereochemical consequences of the in-line and adjacent mechanisms for the first part of ribonuclease action discussed are compared in Figure 6.34. A similar diagram may be constructed for the second part of ribonuclease action.

RATE EFFECTS. It has been observed (*203*) that uridine-2′,3′-cyclic phosphonate **CXIII** is hydrolyzed by ribonuclease A about 1900 times slower than the rate for the corresponding phosphate **CXIV** at pH 6. By way of contrast, the rates of nonenzymic hydrolysis (*204, 205*) are comparable for the two compounds in base or in acid but at different levels than in the enzyme-catalyzed reaction. In view of the in-line mechanism proposed (*198*) for the second part of ribonuclease action on the related phosphorothioate, **CXII**, it seems likely that either Phosphonate **CXIII** is bound by the enzyme, with an orientation that precludes efficient catalysis or the pH vs. rate optimum of the phosphonate differs significantly from that of the phosphate.

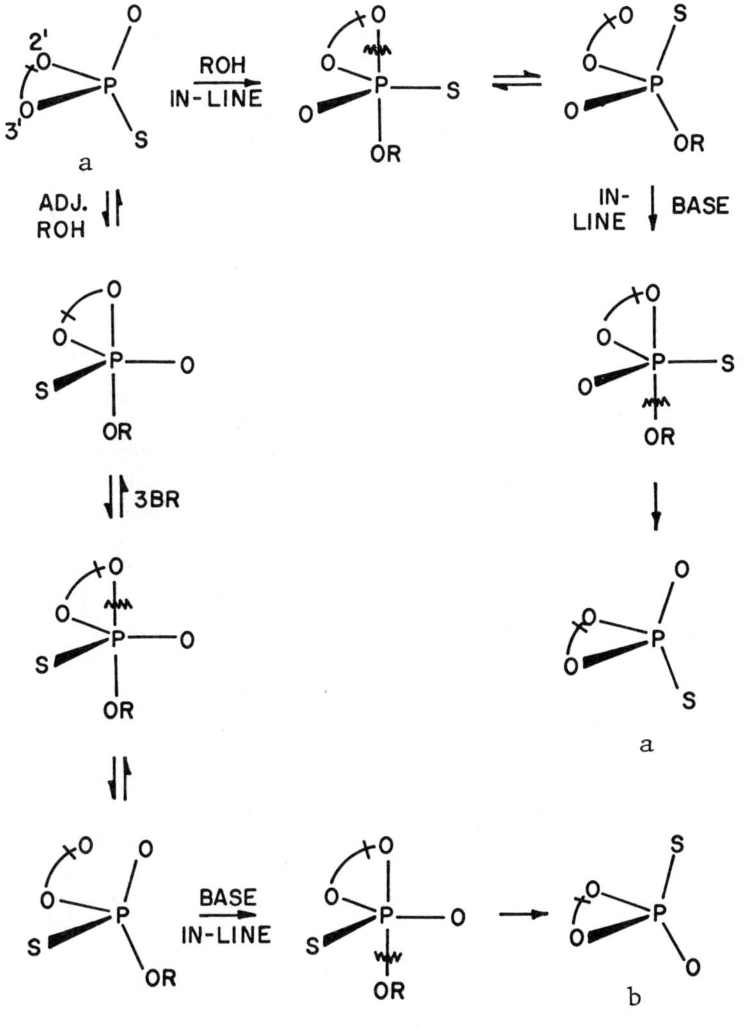

International Journal of Peptide and Protein Research

Figure 2.34. Schematic diagram comparing the stereochemical consequences of the in-line and adjacent mechanisms (207) for the model study (202) of the first part of RNase action outlined in Figure 2.33. The nonenzymatic, base-catalyzed reaction is known to proceed in-line.

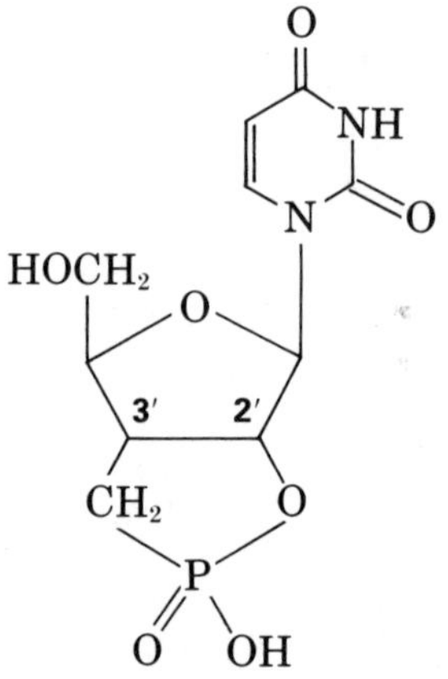

CXIII 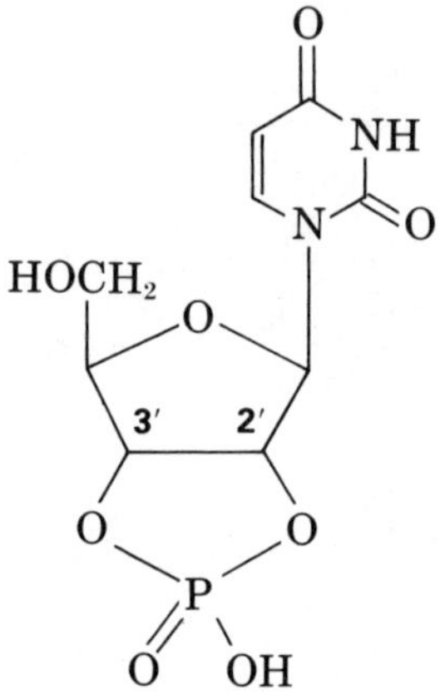   CXIV

SULFUR EXCHANGE. A further interesting observation concerns the hyrolysis of **CXII**. Whereas both isomers exchanged sulfur into the solvent in nonenzymic acid hydrolysis (*206*), no sulfur was detected in the non-enzymic base hydrolysis or hydrolysis catalyzed by ribonuclease (*205*).

These results are interpretable in terms of previous observations (cf. p 110). Pseudorotation would allow the sulfur atom to occupy a leaving axial position. By analogy with the hydrolysis of methylethylene phosphate (pp 93 and 100), the occurrence of pseudorotation is expected in the acid hydrolysis of **CXII** but not in base hydrolysis. The pH of the enzymic hydrolysis seems to be close enough to that of the nonenzymic base hydrolysis to be explainable in terms of inhibition of pseudorotation and not by any special constraints imposed by the enzyme on the substrate. The latter argument agrees with the finding (*205*) that the two isomers are bound to different extents by the enzyme. The different binding might be expected to result in different sulfur exchange rates if pseudorotation were a necessary process. Thus, additional support is provided for the in-line mechanism.

Although the model studies by Usher and co-workers (*198, 202*) support an in-line mechanism for both steps of RNase action proceeding via a trigonal

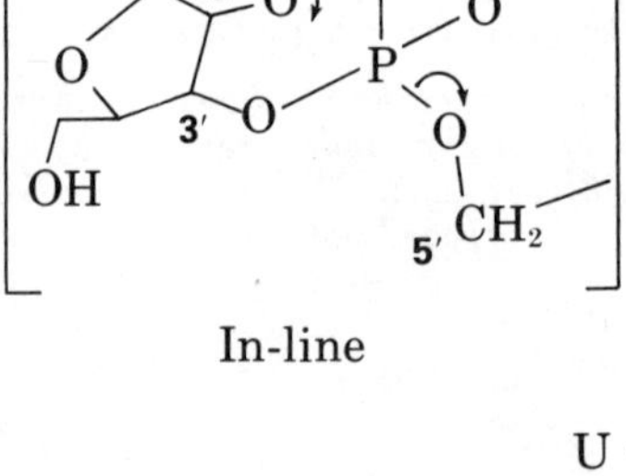

In-line 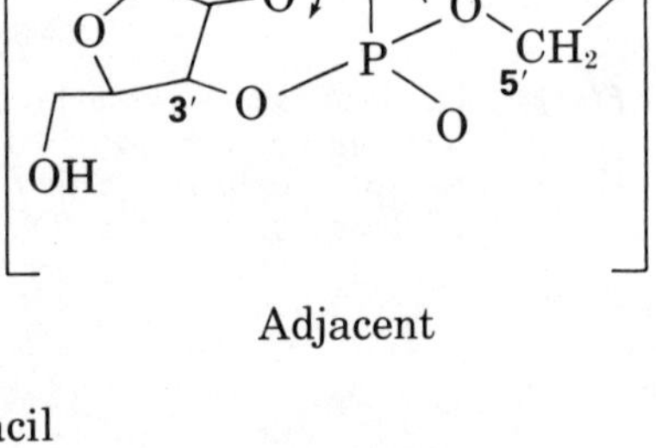   Adjacent

U = uracil

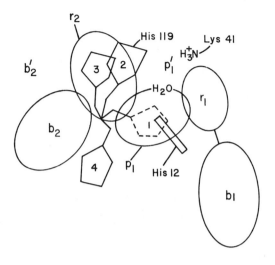

Clarendon

Figure 2.35. Approximate locations of principal components of dinucleoside phosphate substrates inferred to lie in the active site of RNase (193)

bipyramidal intermediate, the results of these studies are also consistent with the presence of square pyramidal intermediates formed during trans basal displacements (an in-line process) (*207*). If the displacement were to occur at cis basal positions of the square pyramid, the reaction would be classified as an adjacent one (cf. p. 181). However, the results obtained (*198, 202*) are not consistent with this process. A detailed comparison of the model study (*198, 202*) of the first step of RNase action proceeding by both in-line and adjacent mechanisms for trigonal bipyramidal and square pyramidal intermediates is given elsewhere (*207*). Later on, we show that considerable distortion toward the inherently less stable square pyramid does seem to exist.

In the mechanism detailed by Roberts et al. (*194*) (Figure 2.31), which incorporates some of the basic features of the mechanism proposed by Rabin and co-workers (*208*), separate roles are assigned to histidine 12 and histidine 119. These roles have gained definition principally by studies of RNase inhibitor complexes using NMR spectroscopy and x-ray diffraction (*185*). As summarized by Richards and Wyckoff (*185, 193*), histidine 119 is found in one of four different positions, histidine 12 is behind the phosphate group and lysine 41 appears at the upper right (Figure 2.35).

A bound solvent molecule also is indicated at the active site. A bound sulfate excludes histidine 119 from Position 1 in the native enzyme, and in this case Position 2 probably is occupied by a solvent molecule. Position 3 is slightly stabilized by cytidine 3′-monophosphate (3′ CMP). However when the adenine nucleoside is present, histidine 119 occupies Position 4. The pyrimidine nucleotides, 3′ CMP and uridine 3′ monophosphate (3′ UMP), occupy

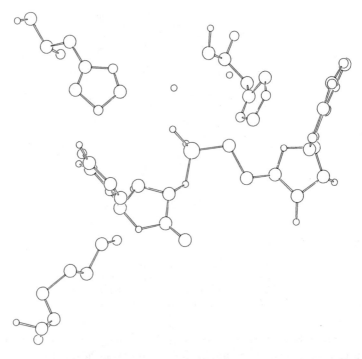

Figure 2.36. ORTEP plot from X-ray coordinates of RNase S–UpcA substrate geometry (193). The uracil base is attached to the ribose moiety on the left and the adenine base is attached to the ribose on the right (see Figure 2.37). Positioned at the active site are lysine 41 (lower left), histidine 12 (upper left), and histidine 119 (upper right). Two water molecules are indicated by unattached open circles.

b_1, r_1, and p_1. Purine nucleotides, adenosine 3'-monophosphate (3' AMP), 5' AMP, and 3' 5' A > p, occupy b_2, and r_2, although b_2, may be any base, while the dinucleotide analogue UpcA occupies both sets of positions. UpcA corresponds to the usual dinucleoside phosphate UpA except the 5' oxygen atom of the ribose attached to the adenine base has been replaced by a methylene group to render it inert to RNase action. A view of this dinucleotide substrate in place at the active site of ribonuclease S is shown in Figure 2.36. A ring nitrogen atom of histidine 12 is near the 2' oxygen atom of the ribose residue r_1, while a nitrogen atom of histidine 119 is near what would be the 5' oxygen atom of the ribose, r_2, bound to the adenine base.

Position p_1' (Figure 2.35) is an inferred position of the cyclic phosphate formed in the first step of the enzymatic catalysis and brings the phosphate close to lysine 41. In arriving at p_1' minimal motion of the ribose is assumed. Position b_2' is an alternate position for the second base. UMP appears to bind weakly at this site.

In the mechanism (*194*) outlined in Figure 2.31, several of whose steps may be concerted, we note the following features. Action is initiated by a proton transfer from the 2'-OH to histidine 12, thus activating the 2' oxygen for incipient formation of a pentacoordinate intermediate via axial attack. The nucleophilicity of phosphorus is enhanced by hydrogen bonding of the phosphoryl oxygens to histidine 119 and lysine 41. The latter presumably becomes more effective as the phosphate moves on cyclization from p_1 to p_1' (Figure 2.35). In the proposed monocyclic trigonal bipyramidal intermediate, the phosphate oxygens receive additional negative charge, leading to a strengthening of the hydrogen-bonding interactions with histidine 119 and lysine 41. It is significant that the pK of histidine 119 is sensitive to the position of the phosphate group while that of histidine 12 is not; this suggests that the latter remains protonated on cyclization. Probably the leaving OR⁻ group receives its proton from histidine 119 in a concerted fashion, although the details of the transfer are uncertain.

The second step—hydrolysis of the cyclic intermediate—is essentially the reverse of the first step with a water molecule in place of the departing nucleotide. The two- to threefold order of magnitude rate reduction for the hydrolysis step compared with the transesterification can be related to a number of differences in the two steps. First, the nature of the base is an important feature that causes variations in measured rates (especially in the first step) and is effective in differentiating the two steps (*209, 210, 212*). For example, the nucleotides CpA and UpA undergo enzymatic catalysis (first step) approximately 1000 times as fast as C > p and U > p, respectively (second step). For each step of the catalytic reaction, cytosine is more effective than uracil. For the first step, the order of decreasing rate as a function of base in the second position (b_2) is adenine > guanine > cytosine > uracil. Other rate data are in Tables 2.15 (*210*) and 2.16 (*212*). Neither A > p nor G > p bind to RNase and, thus, do not act as substrates or inhibitors.

Base interactions caused by both hydrogen bonding to enzyme residues and steric fit, particularly at the base pocket b_2 (Figure 2.35), should be

**Table 2.15.   Steady-State Rate Constants[a] (at pH of 7) for
the First Step of Bovine Pancreatic Ribonuclease Action
(Ester Cleavage) (210)**

Entry	Phosphate Substrate[b]		k_3 (sec^{-1})[c]
1	Cytidylyl 3′,5′-adenosine[d]	(CpA)	2350
2	Cytidylyl 3′,5′-2′-deoxyadenosine	(CpdA)	2350
3	Cytidylyl 3′,5′-guanosine	(CpG)	220
4	Cytidylyl 3′,5′-purine-9-riboside	(Cp purine-9-riboside)	600
5	Cytidylyl 3′,5′-N^6-methyladenosine	(Cpm⁶A)	90
6	Cytidylyl 3′,5′-N^6-dimethyladenosine	(Cpm₂⁶A)	40
7	Cytidylyl 3′,5′-3-isoadenosine	(Cp3isoA)	400
8	Cytidylyl 3′,5′-cytidine	(CpC)	160
9	Cytidylyl 3′,5′-uridine	(CpU)	27
10	Cytidylyl 3′,5′-N^3-methyluridine	(Cpm³U)	18
11	Cytidylyl 3′,5′-thymidine	(CpT)	15
12	Cytidylyl 3′,8′-2′,3′-isopropylidene-5′-deoxy-5′-thioethyladenosine	—	28
13	Cytidylyl 3′,6′-9-hexyladenine	—	60
14	Cytidylyl 3′,5′-9-pentyladenine	—	30
15	Cytidylyl 3′,4′-9-butyladenine	—	50
16	Cytidylyl 3′,3′-9-propyladenine	—	3
17	Cytidylyl 3′,3′-2-propylbenzimidazole	—	3
18	Cytidylyl 3′,3′-2′-deoxyadenosine	—	<3
19	Uridylyl 3′,5′-adenosine	(UpA)	1000
20	Uridylyl 3′,5′-adenosine-N^1-oxide	(UpO¹A)	14
21	Uridylyl 3′,5′-guanosine	(UpG)	69
22	Uridylyl 3′,5′-cytidine	(UpC)	26
23	Uridylyl 3′,5′-cytidine-N^3-oxide	(UpO³C)	2
24	Uridylyl 3′,5′-N^3-methylcytidine	(Upm³C)	20
25	Uridylyl 3′,5′-uridine	(UpU)	11

Table 2.15.   Continued

Nucleoside formula[e] ($r_2 b_2$)

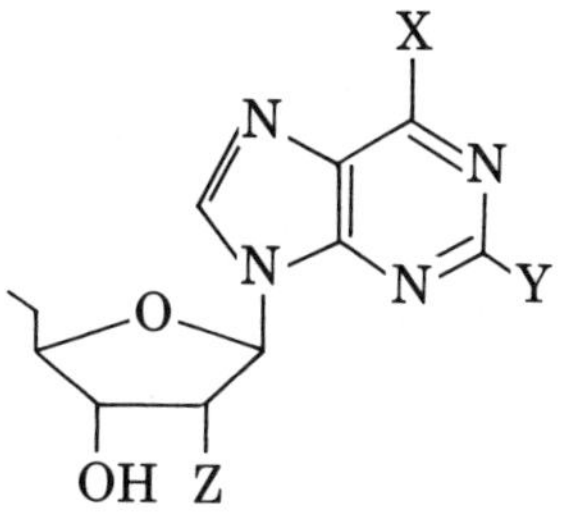

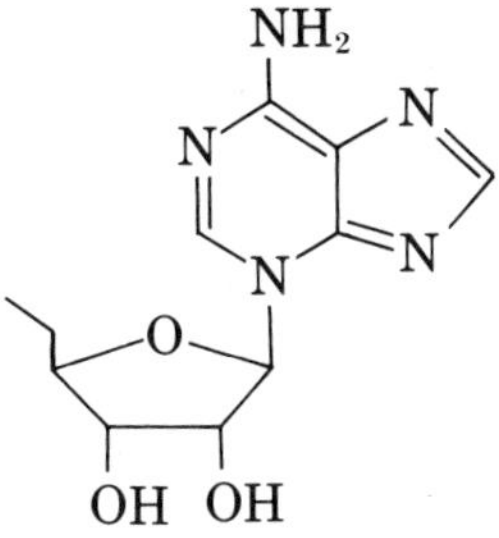

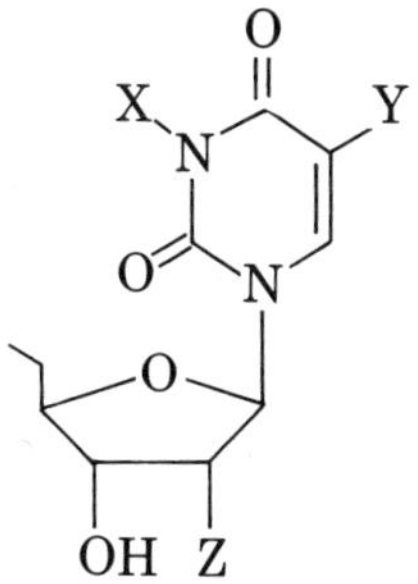

	X	Y	Z			X	Y	Z
1	NH_2	H	OH	**7**	**9**	H	Ḣ	OH
2	NH_2	H	H		**10**	CH_3	H	OH
3	OH	NH_2	OH		**11**	H	CH_3	H
4	H	H	OH					
5	$NHCH_3$	H	OH					
6	$N(CH_3)_2$	H	OH					

$-CH_2CH_2S-$

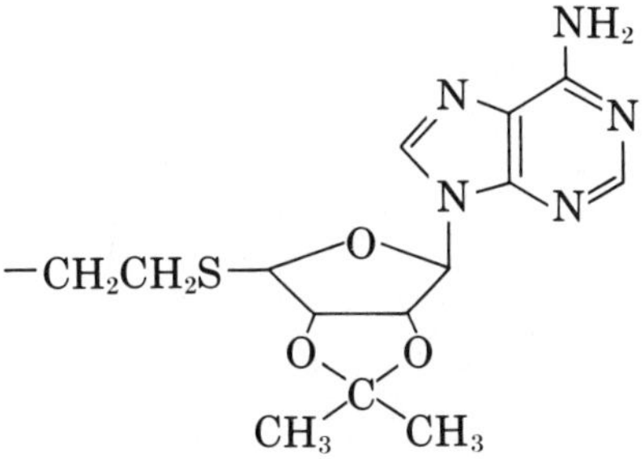

12

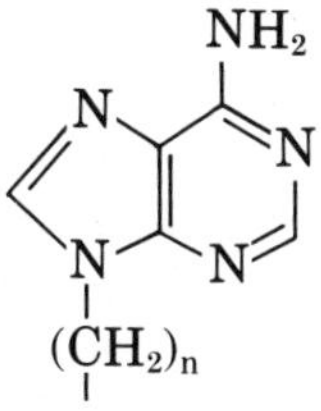

13	$n = 6$
14	$= 5$
15	$= 4$
16	$= 3$

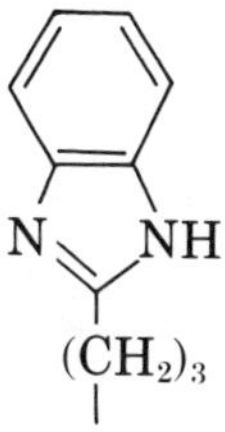

17

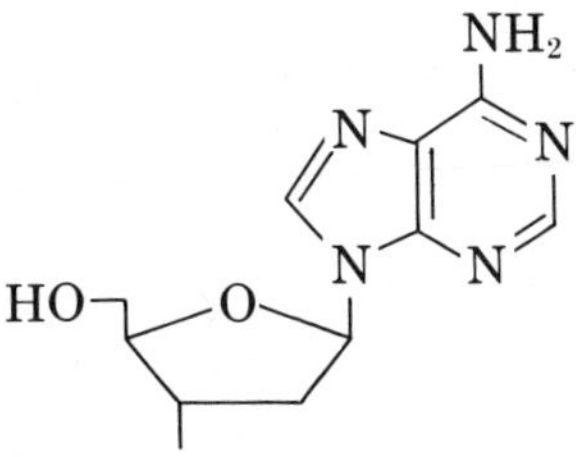

18

Table 2.15. Continued

Nucleoside formula[e] (r_2b_2)

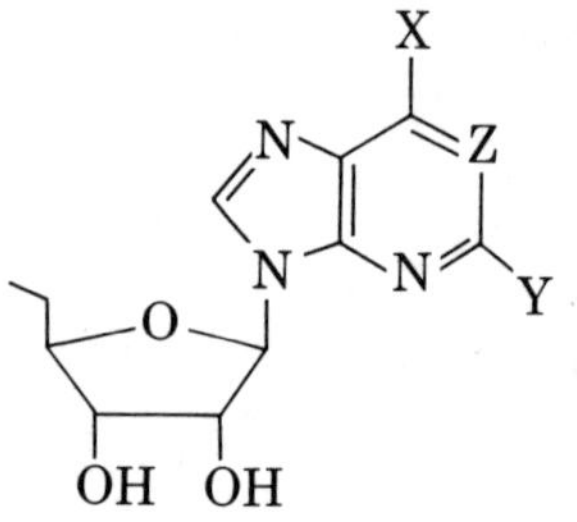

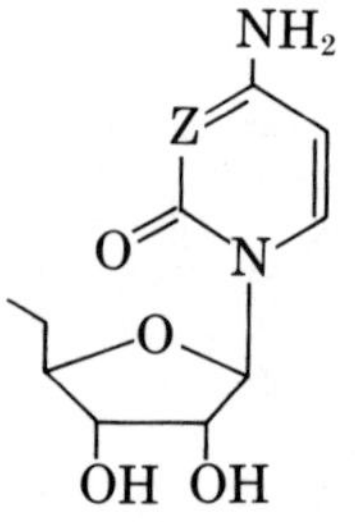

	X	*Y*	*Z*
19	NH_2	H	N
20	NH_2	H	N^+-O^-
21	OH	NH_2	N

	Z
22	N
23	N^+-O^-
24	N^+-CH_3

[a] Data from Ref *210*.

[b] Solvent was 0.1*M* dimethylglutaric acid; NaOH + NaCl to give $I = 0.2M$, pH 7.0 at 25°C.

[c] The constant k_3 measures the rate of dissociation of the enzyme–substrate complex (ES) according to:

$$E + S \underset{k_2}{\overset{k_1}{\rightleftharpoons}} ES \overset{k_3}{\rightarrow} E + P$$

and is called the turnover number of the enzyme. This is the number of substrate molecules converted into Product P per unit time when the enzyme is saturated with the substrate. See Ref *211* for further discussion of enzyme kinetics.

[d] The term phosphate has been omitted in the name of the entries.

[e] Nucleoside indentification (r_2b_2) in the following dinucleoside phosphate formulation, where b_1 is cytosine for Entries **1–18** and uracil for Entries **19–24**. The nucleoside formula (r_2b_2) for uridine and cytidine is shown under Entries **9** and **22**, respectively.

European Journal of Biochemistry

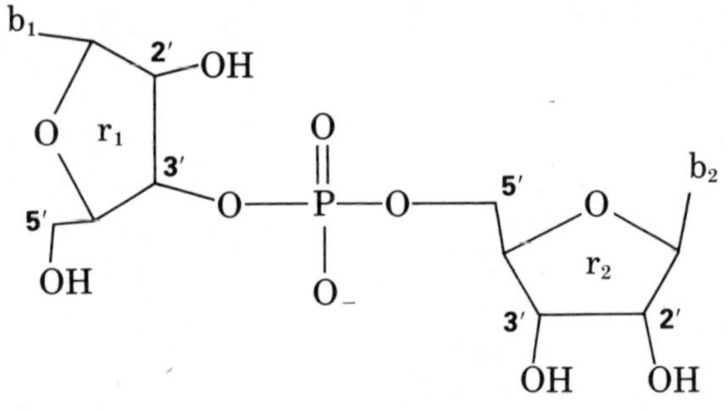

Table 2.16. Steady-State Rate Constants[a] (at pH of 7) for the Hydrolysis of Cyclic Phosphates by Bovine Pancreatic Ribonuclease (Second Step) (212)

Entry	Cyclic Phosphate Substrate[b]		k_3 (sec⁻¹)[c]
1	Cytidine 2′,3′-phosphate	C > p	5.5
2	4-N-Acetylcytidine 2′,3′-phosphate	ac⁴C > p	0.5
3	4-N-Methylcytidine 2′,3′-phosphate	m⁴C > p	1.9
4	4-N-Dimethylcytidine 2′,3′-phosphate	m₂⁴C > p	0.2
5	Uridine 2′,3′-phosphate	U > p	1.4
6	3-Methyluridine 2′,3′-phosphate	m³U > p	No inhibition
7	5,6-Dihydrouridine 2′,3′-phosphate	H₂U > p	0.5
8	5-Chlorouridine 2′,3′-phosphate	Cl⁵U > p	1.7
9	5-Bromouridine 2′,3′-phosphate	Br⁵U > p	1.3
10	5-Iodouridine 2′,3′-phosphate	I⁵U > p	1.2
11	4-Thiouridine 2′,3′-phosphate	s⁴U > p	5.2
12	4-Methylthiouridine 2′,3′-phosphate	ms⁴U > p	1.9
13	Pseudouridine 2′,3′-phosphate	ψU > p	0.3
14	1-Methylpseudouridine 2′,3′-phosphate	m¹ψU > p	1.6
15	3-Methylpseudouridine 2′,3′-phosphate	m³ψU > p	—
16	1,3-Dimethylpseudouridine 2′,3′-phosphate	m₂¹,³ψU > p	No inhibition
17	N-3-Uric acid riboside 2′,3′-phosphate	N-3-Uric > p	0.5
18	N-9-Uric acid riboside 2′,3′-phosphate	N-9-Uric > p	0.1
19	8-Bromoguanosine 2′,3′-phosphate	8-BrG > p	—
20	8-Methylthioguanosine 2′,3′-phosphate	8-CH₃SG > p	—
21	8-Oxyguanosine 2′,3′-phosphate	8-OxyG > p	1.3
22	1(β-D-Ribofuranosyl)-2-pyridone 2′,3′-phosphate	—	0.2
23	1(β-D-Ribofuranosyl)-3-methyl-2-pyridone 2′,3′-phosphate	—	—
24	1(β-D-Ribofuranosyl)-3,6-dioxypyridazine 2′,3′-phosphate	—	0.2

Base Formula[d] (b₁)

	1	R = H, R′ = H
	2	R = H, R′ = COCH₃
	3	R = H, R′ = CH₃
	4	R = CH₃, R′ = CH₃

13	R = H, R′ = H
14	R = H, R′ = CH₃
15	R = CH₃, R′ = H
16	R = CH₃, R′ = CH₃

Table 2.16.   Continued

Base Formula[d] (b₁)

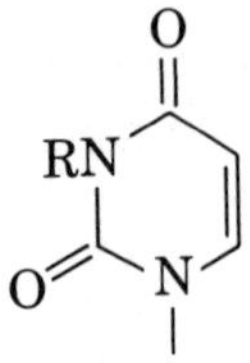

5 R = H
6 R = CH₃

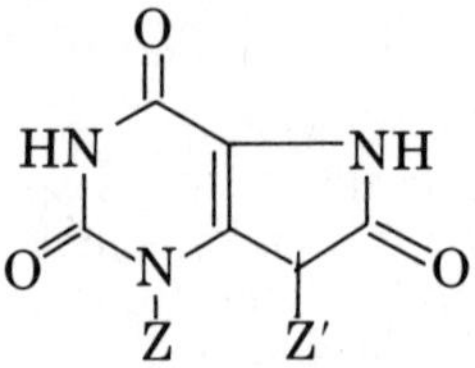

17 Z = Rib-P
       Z′ = H
18 Z = H
       Z′ = Rib-P

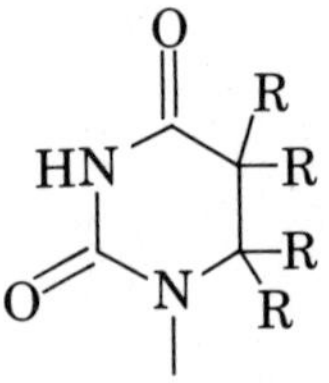

7 R = H

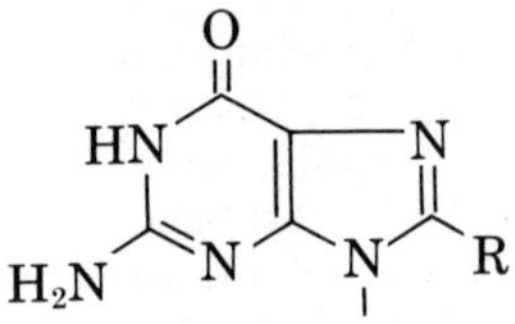

19 R = Br
20 R = SCH₃
21 R = OH

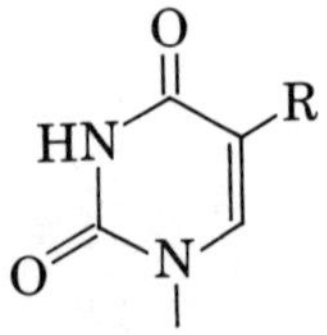

8 R = Cl
9 R = Br
10 R = I

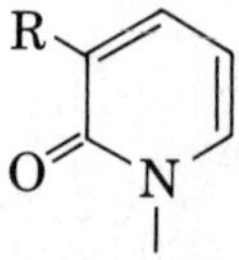

22 R = H
23 R = CH₃

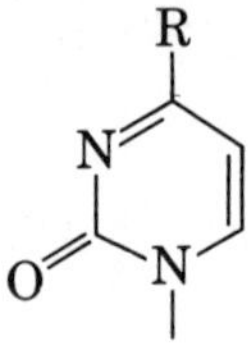

11 R = SH
12 R = SCH₃

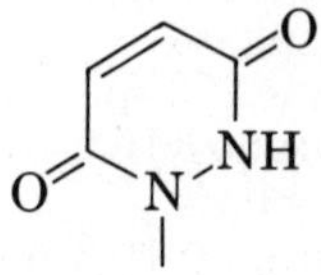

24

[a] Data from Ref *212*.
[b] Same solvent conditions as stated in Footnote *b* to Table 2.15.
[c] *See* Footnote *c* to Table 2.15.
[d] Base identification (b₁) in the cyclic nucleotide substrate:

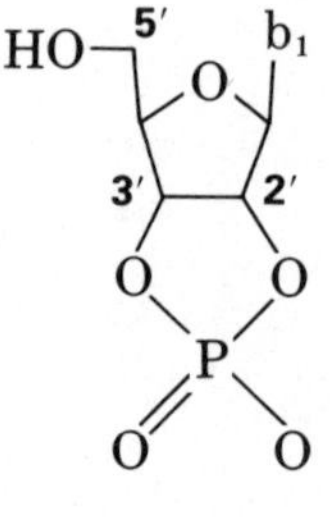

European Journal of Biochemistry

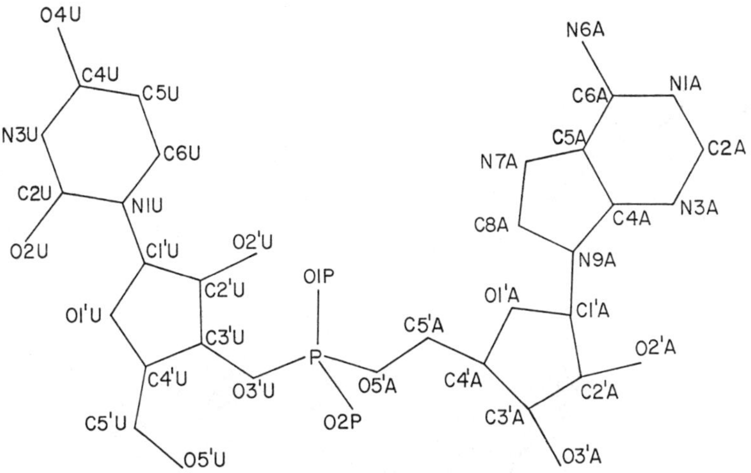

Figure 2.37. Conventional numbering scheme for UpcA used in conjunction with bond parameters listed in Table 2.17

important in accounting for rate variations encountered in the first step of RNase action. Specific hydrogen-bonding interactions have been formulated for the binding of uracil to RNase (*see* p 191) based on the x-ray study of the RNase S–UpcA complex (*193, 213*).

It is instructive that the presence of the 2-keto function, which the purines lack, appears necessary for substrate binding during the hydrolysis step. Consistent with known hydrogen-bond interactions, two such bonds present in a closed-ring system (as with threonine 45 here) are stronger than two individual hydrogen bonds (*214*). Perhaps the inability of A > p and G > p to bind to RNase is associated with the weaker hydrogen-bonding system expected at the base pocket (b_1 of Figure 2.35) for these cyclic phosphates compared with those for U > p and C > p. Presumably, similar hydrogen-bonding interactions are present at the base site, b_2.

In general, the interaction of nucleosides is stronger than that of either ribose or free bases, and nucleotides bind more strongly than nucleosides (*185*). If the effectiveness of catalysis is related directly to the ability of the enzyme to position the substrate toward its transition state geometry (*184*), then we must assume that a base, which is properly oriented in its enzyme pocket, aids in the attainment of this configuration. It is significant that the uracil and adenine bases, as seen in the RNase S–UpcA complex (Figure 2.36), are held nearly planar to one another. Also, the ribose units are nearly planar to each other but orthogonal to the bases. The structural formula for the UpcA substrate is given in Figure 2.37. Because of the particular location of the enzyme pockets, the phosphate moiety is positioned correctly for an in-line attack by the 2′-OH of ribose r_1, leading to a trigonal bipyramidal

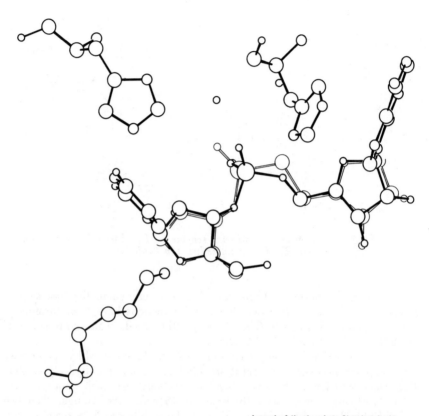

Figure 2.38. ORTEP drawing of the minimum energy structure computed for the UpA substrate bound by RNase, superimposed on the X-ray coordinates of UpcA (lighter lines) (217)

transition state. As a result of this positioning, the 2′-OH of uridine is only 3.31 Å from the phosphorus atom (*193*). This compares with 3.81 and 4.11 Å for the two different molecules of naturally occurring uridylyl-3′, 5′-adenosine UpA (*215*). The closeness of the 2′-OH ensures rapid formation of the 2′, 3′-cyclic phosphate in the first part of RNase action. Conversely, use of a poorer fitting base, especially at b_2, should make attainment of the transition-state geometry more difficult and result in a slower rate for transphosphorylation. In this regard, the water molecule, viewed as the attacking agent in the hydrolysis step, lacks the more complex enzyme interactions associated with nucleosides. Without this supportive effect, the enzyme is less able to promote the formation of the expected trigonal bipyramidal transition state, leading to ring opening. Although not substantiated, this discussion provides a rationale for the much slower reaction rate of the second part of RNase action and, perhaps for some of the rate variations obtained with the different bases (listed in Table 2.15), for the first part of RNase action.

A great deal is known now about the active site interactions for RNase, particularly those present in the enzyme–substrate complex (*185,193,194,216*). However, similarly detailed information is lacking about transition states and intermediates appearing along the mechanistic course leading to the hydrolysis of a ribonucleic acid. As discussed, inferential evidence favors trigonal bipyramidal transition states and the formation of a cyclic intermediate. Since a great deal also is know about phosphorane structures (Volume I, p 11) and these model the proposed transition states in RNase action to varying degrees, this information should be advantageous in ascertaining how a pentacoordinated conformation is modified by enzyme constraints. Hence, it might be possible to establish a reliable estimate of the geometric pathway for ribonuclease action.

An attempt in this direction has been made by applying a molecular mechanics model, which was designed to reproduce the structures of phosphorus compounds by computer simulation (*see* Volume I, p. 293), to the RNase system (*217*). With sufficient parameterization available from structural studies of cyclic and acyclic derivatives to establish a set of "strainless" bond lengths and bond angles and associated force constants defined by vibrational studies, a molecular mechanics approach using an energy minimization scheme seems well suited to investigate structural perturbations caused by enzyme constraints. Results of x-ray studies of nucleotides also contributed to the parameterization; for example, results from a crystal study of cytidine 2′, 3′-cyclic phosphate (C > p) (*218*), an intermediate in the first step of RNase action (*see* Table 2.16), and from a crystal study of the naturally occurring dinucleoside, uridylyl 3′, 5′-adenosine (UpA) (*215*).

It has been established that the structures of cyclic pentacoordinated phosphorus compounds obtained by x-ray diffraction show distortions from idealized tirgonal bipyramidal symmetry, and that these distortions closely follow the local C_{2v} constraint of the Berry intramolecular exchange coordinate (Volume I, p 34). Further, the molecular mechanics model can reproduce the major structural deviations encountered (Volume I, p 297). Hence, it is expected that if active site interactions are reasonably well defined for the proposed pentacoordinated transition states in RNase action, the molecular mechanics approach should yield probable geometrical representations.

In simulating the first step of RNase action, the coordinates from the single crystal x-ray study of RNase S–UpcA provided the initial active-site geometry (*193*). Replacement of the methylene group with an oxygen atom gave UpA. Subsequent conformational minimization led to refinement of the enzyme–substrate complex (Figure 2.38). Further, by rotating histidine 119 from its position in the RNase–UpA complex (Figure 2.38), a lower-energy structure for the initial RNase–UpA complex is obtained. The geometries at phosphorus for the RNase–UpcA and computed RNase–UpA complexes are compared with that for native UpA in Table 2.17. Compared

Table 2.17.   Computed Geometries at Phosphorus

	Native UpA X-Ray[b]	Enzyme Bound	
		UpcA X-Ray[c]	UpA Computed[d]
Bond Lengths (Å)			
$P-O_{3'U}$	1.62	1.61	1.61
$P-O_1$	1.47	1.46	1.48
$P-O_2$	1.48	1.66	1.47
$P-O_{5'A}$	1.60	1.85 (P–C)	1.61
$P-O_{2'U}$	4.11[e] nb	3.31 nb	3.51 nb
Bond Angles (degrees)			
$O_{3'U}-P-O_1$	108.9	113.1	107.1
$O_{3'U}-P-O_2$	110.0	110.1	112.2
$O_{3'U}-P-O_{5'A}$	100.6	106.0	102.0
O_1-P-O_2	118.8	108.3	111.9
$O_1-P-O_{5'A}$	106.6	106.0	110.2
$O_2-P-O_{5'A}$	110.7	113.4	112.9
$O_{2'U}-P-O_{5'A}$	—	154.9 nb	141.0 nb
$O_{2'U}-P-O_{3'U}$	—	49.0 nb	47.0 nb
$O_{2'U}-P-O_1$	—	86.0 nb	69.0 nb
$O_{2'U}-P-O_2$	—	82.3 nb	102.3 nb

[a] Refer to Figure 2.37 for atom labeling.
[b] From Ref *215*. There are two different molecules of UpA per unit cell. The bond distances and angles listed here are those for UpA2. The corresponding parameters for UpA1 differ only slightly from those for UpA2.

with the geometry of the latter, the computed geometry for bound UpA, like that from the x-ray study of the UpcA complex, shows the substrate poised for incipient nucleophilic attack by the 2'-OH group. The P–O(2')U bond, which is 4.11 and 3.81 Å in the two forms of naturally occurring UpA (*215*), is calculated to be 3.51 Å in the RNase–UpA complex.

By requiring that the phosphorus atom move in a straight line, energy minimization leads to a transition-state conformation (Figure 2.39) with P–O(2')U and P–O(5')A distances of approximately 2.2 Å in contrast to 3.5 Å for PO(2')U and 1.6 Å for the P–O(5')A distances computed for the enzyme–UpA complex. The geometry at phosphorus is intermediate between an idealized trigonal bipyramid and square pyramid (Table 2.17). Continuing further along the reaction coordinate results in the formation of the cyclic intermediate, uridine 2', 3'-cyclic phosphate (U>p). The latter minimum-energy structure is included in Table 2.17 and portrayed in Figure 2.39. In the process leading to the cyclic phosphate, a near 2-Å movement of the

for the First Step of RNase Action on UpA[c]

Transition State UpA Computed[d]	*Cyclic Intermediate UpA Computed*[d]
Bond Lengths (Å)	
1.61	1.61
1.47	1.48
1.47	1.48
2.16	3.55 nb
2.16	1.61
Bond Angles (degrees)	
112.7	109.1
127.4	111.8
71.6	54.4 nb
119.8	114.5
103.9	126.7 nb
87.1	57.6 nb
150.2	123.1 nb
78.8	103.0
104.3	109.6
87.3 nb	108.2

[c] From Ref *193*. The atom labeled $O_{5'A}$ is a C atom in UpcA.
[d] From Ref *217*.
[e] In UpA1, this nonbonded (nb) $P-O_{2'U}$ distance is 3.81 Å.

phosphorus atom from its initial position in the RNase–UpA complex occurs. Accompanying this movement, lysine 41 comes into effective interaction with the phosphate, whereas only histidine 119 and histidine 12 are so positioned in the initial substrate complex.

All of the features obtained from the computed reaction course based on parameterization of a molecular mechanics program agree with conjectures previously advanced for this well-defined enzyme system. As more structural information becomes available about active-site environments in other phosphoryl and nucleotidyl transfer enzymes, application of computer-based molecular mechanics models should provide real insight into mechanistic features unattainable by other means. Other enzyme systems whose mechanisms seem amenable to this technique are outlined in the next two sections.

DNA Replication and RNA Transcription. An important area in which application of mechanistic considerations based on pentacoordinate principles appears manifest concerns the function of polymerases in DNA

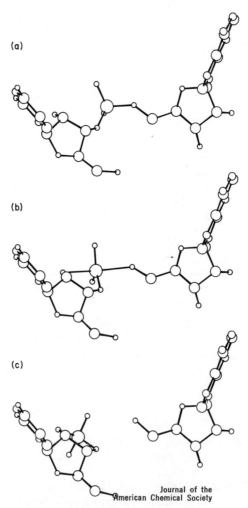

(a)

(b)

(c)

*Figure 2.39. Minimum-energy con-
formations computed for UpA bound
by RNase (217). (a) Initial substrate
conformation, (b) five-coordinated
transition state, (c) 2', 3'-cyclic intermediate. In obtaining these representations, histidine 119 was rotated
to a more favorable orientation relative to that in Figure 2.36.*

replication and repair, on the one hand, and RNA transcription, on the other. Although the topic is vast, we concern ourselves only with specific aspects of DNA replication dealing with the proposed five-coordinated intermediate and show analogies between this process and RNA transcription. For readers interested in extending their knowledge in this area, several sources are available (*219–226*) that cover many facets for these and related enzyme systems. Even at this stage in the development of models for enzyme action involving the ubiquitous phosphate entity, pentacoordinate stereochemistries and their associated mechanistic criteria seem destined to become the dominant influence in describing active-site formulations.

The overall process of DNA replication can be discussed in terms of initiation, chain elongation, and termination (for an excellent introduction, *see* Ref 227). Each process represents a complex series of events in itself. In the process that we wish to discuss—i.e., chain elongation—both parental strands of DNA are duplicated simultaneously as the double helix unwinds. The parent strands act as templates for the formation of new complementary strands. Various DNA polymerases are involved as chain elongation proceeds in the $5' \rightarrow 3'$ direction. A portion of a DNA chain is shown here in abbreviated notation:

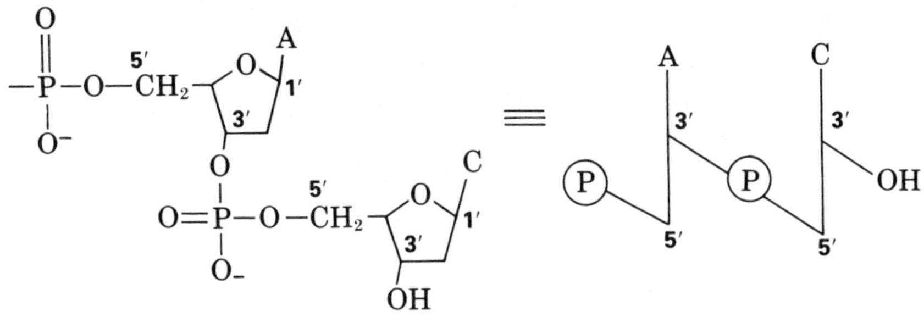

A represents adenine, and C is cytosine. Growth in the $5' \rightarrow 3'$ direction adds successive nucleotides to the right, attaching each to the unlinked $3'$-OH group. Chain elongation proceeds at a rate of about 1000 nucleotides per minute per molecule of DNA polymerase I.

To understand the process of chain elongation, remember that the two strands of the double helix run in opposite directions (Figure 2.40). Further, since the important polymerases I, II, and III found in *E. coli*, where much of the work on the mechanism of chain elongation has been done (*219*), synthesize DNA only in the $5' \rightarrow 3'$ direction, one must conclude that DNA replication is a discontinuous process. In other words, the implied event in Figure 2.41—the simultaneous continuous syntheses of each parental strand as the original DNA helix unwinds—is incorrect since it shows only one strand being duplicated in the correct $5' \rightarrow 3'$ direction, whereas template synthesis of the other strand is shown proceeding in the $3' \rightarrow 5'$ direction. It has been established that short segments (Okazaki fragments (*228, 229*)) of about 1000 nucleotides each are synthesized in the $5' \rightarrow 3'$ direction, and

(a)

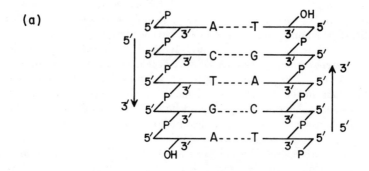

(b)

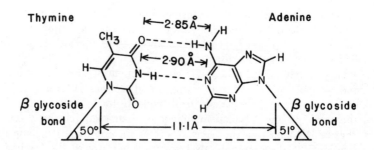

Figure 2.40. (a) Antiparallel orientation of the complementary strands of a DNA segment; (b) hydrogen-bonded base pairs, cytosine–guanine (C–G) and thymine–adenine (T–A).

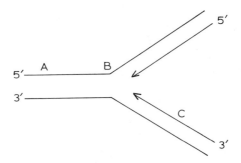

Figure 2.41. Apparent (but incorrect) direction of replication for daughter strand (C), 3′ → 5′, as parental DNA (A) unwinds at the replicating fork (B)

then these nascent segments are joined by a DNA ligase. Accordingly, both daughter strands grow in the 5′ → 3′ direction.

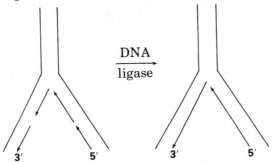

The detailed mechanism of chain elongation has reecived considerable impetus by conformational studies using NMR techniques. Sloan, Loeb, Mildvan, and Feldmann (*230*) measured the NMR rates of the α, β, and γ phosphorus atoms and five protons of thymidine 5′-triphosphate-(dTTP) interacting with paramagnetic Mn(II) at the active site of *E. coli* DNA polymerase I. These data and corresponding data on the binary Mn(II)–dTTP complex resulted in the Mn(II)–proton and Mn(II)–phosphorus atom distances given in Table 2.18.

The results for the ternary complex indicate that Mn(II) coordinates only with the γ-phosphoryl group and that the conformation of the enzyme-bound dTTP is puckered whereas all three phosphorus atoms are coordinated to Mn(II) in the binary complex. Computer calculations (*230*) reveal a unique conformation for the enzyme-bound Mn–dTTP complex with a thymine–deoxyribose torsion angle about the β-glycosidic bond (N_1–C_1') of 90°. The latter, significantly, is indistinguishable from that found in double helical

Table 2.18. Summary of Mn(II)–Proton and Mn(II)–P Bond Distances (Å) in Mn(II)–dTTP and Pol I–Mn(II)–dTTP (230)

	$Mn(II)$–$dTTP$	$Pol\ I$–$Mn(II)$–$dTTP$
αP	3.7 ± 0.3	4.2 ± 0.4
βP	3.4 ± 0.2	4.8 ± 0.3
γP	3.3 ± 0.2	3.2 ± 0.3
H_4'	6.0 ± 0.4	8.8 ± 0.6
H_2'	8.7 ± 0.7	10.8 ± 0.9
H_1'	8.3 ± 0.1	10.3 ± 0.6
H_6	6.9 ± 0.2	9.9 ± 0.7
$-CH_3$	8.7 ± 0.2	10.4 ± 0.7

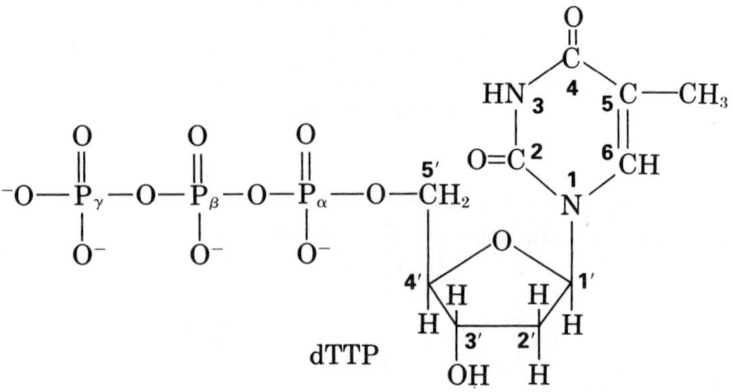

Journal of Biological Chemistry

DNA-B (*231*). This same angle in the absence of enzyme (i.e., in the Mn(II)–dTTP complex) is 40°. The enzyme-bound complex illustrated in Figure 2.42 suggests (*230*) that DNA polymerase adjusts the conformation of dTTP to facilitate correct Waston–Crick base pairing for nucleotidyl incorporation and, hence, contributes to the error-preventing mechanism observed for DNA polymerases (*232*). The error rate for incorporating mismatched nucleotides in the related enzyme-bound magnesium complex is about 1 in 10^5 polymerization events. The Pol I–Mn(II)–dTTP complex, however, is more prone to mistakes (*233*).

With a computer, the optimum conformation of Pol I–Mn(II)–dTTP (Figure 2.42a) was superimposed onto that of a thymidylate residue of DNA. The latter residue serves as the primer terminus (*234*). From the geometry obtained, a mechanism of chain elongation is suggested (*230*), see Figure 2.43. Nucleophilic attack by the 3'-OH group of the primer terminus on the α-phosphorus atom of dTTP is indicated to proceed in-line (axial attack). Adjacent attack in a trigonal bipyramidal intermediate (p 181) is rendered unlikely because of steric interaction between Mn(II) and the H_5' protons of dTTP. Direct coordination of the γ-phosphorus atom by Mn(II) implies activation of the leaving pyrophosphate group while the puckered triphosphate arrange-

(a)   (b)

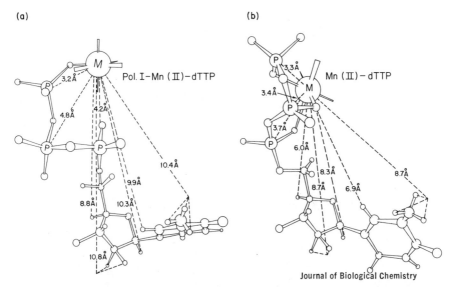

Pol. I–Mn (II)–dTTP

Mn (II) – dTTP

Figure 2.42. Optimum conformations (230) of (a) the ternary DNA poly-merase I–Mn(II)–deoxythymidine 5'-triphosphate complex (Pol I–Mn(II)–dTTP) and (b) the binary Mn(II)–dTTP complex obtained as a best fit of the data in Table 2.18 with the aid of a computer analysis.

ment suggests a possible strain on the α-phosphorus–oxygen bond undergoing cleavage.

Thus, the biosynthesis of nucleic acids (at least in the Pol I system, like their hydrolytic decomposition, as discussed for RNase action (p 191), fits an in-line mechanism involving a trigonal bipyramidal transition state. The mechanism is summarized in a stepwise fashion in Figure 2.44 (232, 235, 236). Included is a role for tightly bound stoichiometric Zn(II), a further require-ment for DNA polymerase activity (232, 237, 238). It is proposed that Zn(II) coordinates and promotes the nucleophilic character of the 3'-hydroxyl group of deoxyribose at the growing point of the primer strand (236, 238). That Zn(II) plays an important role is shown by the loss of catalytic activity of *E. coli* DNA polymerase I in direct proportion to the amount of Zn(II) removed by prolonged dialysis against *o*-phenanthroline acting as a chelating agent (238). Other DNA polymerases appear to be zinc metalloenzymes as well (232, 239).

We see that important enzyme features, as with RNase action (p 180), facilitate incorporation of the substrate toward its transition state conforma-tion. Both stereochemical positioning and charge effects provide easy access to the proposed five-coordinate intermediate.

RNA is also transcribed from a DNA template. This applies to all cel-lular RNA in *E. coli*: messenger (mRNA), transfer (tRNA), and ribosomal (rRNA) (for an introductory discussion on RNA synthesis, *see* Ref 240).

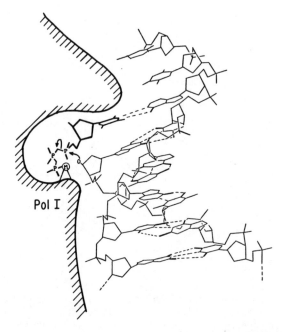

Journal of Biological Chemistry

Figure 2.43. Computer positioning of the optimum conformation of Pol I–Mn(II)–dTTP (from NMR analysis (230)) onto a thymidylate residue of DNA. An in-line mechanism is suggested for chain elongation in the template synthesis of DNA.

There are several similarities between the synthesis of RNA and DNA. Both involve specific polymerases and follow a similar mechanism, adding nucleotidyl phosphates on a chain growing in the 5′ → 3′ direction. Nucleophilic attack by the 3′-hydroxyl group takes place at the terminus of the growing chain on the α-phosphorus atom of the incoming ribonucleoside triphosphate, ATP, GTP, CTP, or UTP (Figure 2.45). Divalent metal ions are involved. As with DNA synthesis, RNA synthesis derives its energy from the subsequent hydrolysis of pyrophosphate.

Unlike DNA synthesis, RNA polymerase does not require a primer and does not have nuclease activities. Further, the DNA template is fully conserved in RNA transcription but semiconserved in DNA replication—i.e., in DNA synthesis parent strands from the double helix unwind on replication and each becomes part of a new double helix containing a daughter strand as a partner whereas local unwinding occurs in RNA synthesis (one strand of the DNA serves as a template, and as synthesis proceeds, the DNA portion that has been copied rewinds).

We discussed the mechanism of chain elongation at the active site of *E. coli* DNA polymerase I (derived from in vitro studies) at a seemingly

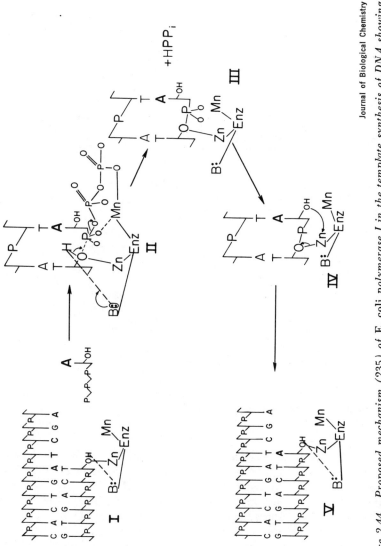

Journal of Biological Chemistry

Figure 2.44. Proposed mechanism (235) of E. coli polymerase I in the template synthesis of DNA showing an in-line attack by the 3'-OH of the primer terminus on the α-phosphorus atom of dATP. PPi is inorganic pyrophosphate, $P_2O_7^{4-}$.

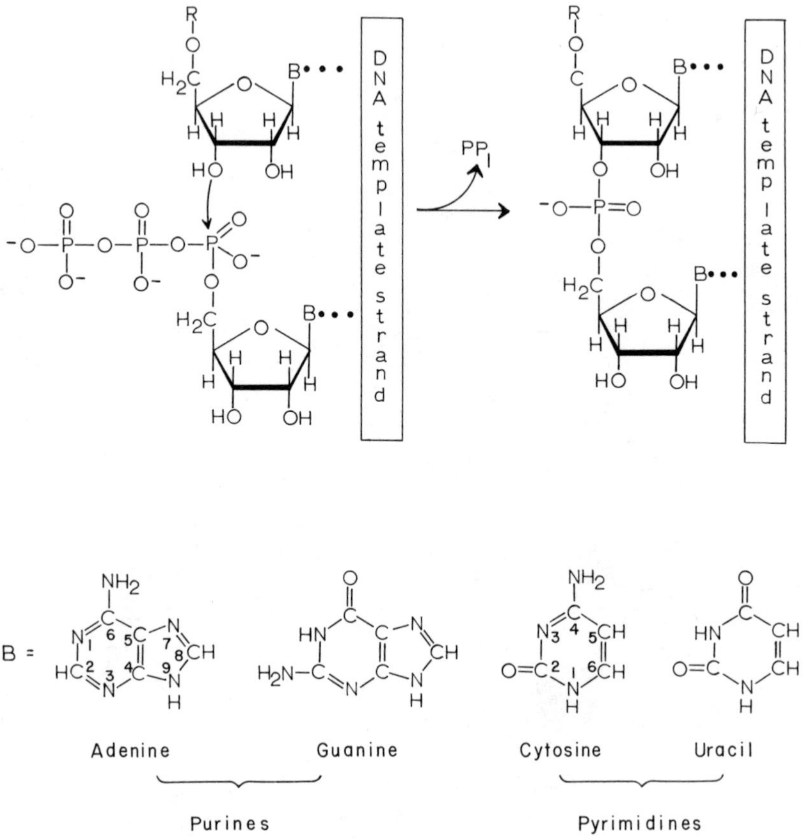

*Figure 2.45. Proposed mechanism of chain elongation of RNA by RNA
polymerase in the presence of a DNA template (221)*

intimate level. It is worthwhile to point out the complexity of the DNA
replication process (*241, 242, 243*) to give a proper focus to the limited but
important aspects just treated.

In *E. coli* alone, more than 20 genes code for proteins that are essential
for DNA replication. Among these are at least two other DNA polymerases.
All three DNA polymerases (Pol I, II, and III) have multiple functions.
Along with polymerization, they show exonuclease activity, both 3′ → 5′ and
5′ → 3′ with Pol I (Figure 2.46) although Pol II and III exhibit only 3′ → 5′
activity. The 3′ → 5′ nuclease activity in Pol I is an editing function, which
results in the removal of mismatched residues at the primer terminus in DNA
synthesis. It is required that the nucleotide that is removed have a free 3′-OH
terminus and not be part of a double helix. This proof reading function
contributes to the high fidelity of DNA replication.

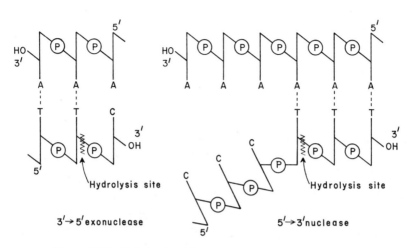

Figure 2.46. Schematic of 3′ → 5′ and 5′ → 3′ nuclease activity

The 5′ → 3′ nuclease activity of Pol I requires a base-paired region. Cleavage can occur at the terminal phosphodiester bond or at a bond up to 10 residues away. Its function in excision repair is enhanced by simultaneous DNA polymerization. Therefore, at no time is a single-stranded gap exposed to attack at the 3′ end by exonucleases. The 5′ → 3′ exonuclease activity is illustrated in Figure 2.47 for the excision of a pyrimidine dimer formed by exposure of DNA to UV light.

Along with these enzymes required for DNA replication, unwinding (*244*) and untwisting (*245, 246*) proteins are necessary, which facilitate the opening of local regions of double helical DNA and relieve the torque and correct the supertwisting that occurs when strands of intact circular DNA are taken apart. Such untwisting enzymes, which apparently operate without an energy source, have been isolated from *E. coli* (*245*) and extracts from mammalian cells (*246*). A mechanism suggested for the latter action (*246*) on superhelical, closed-circular polyoma DNA involves breakage of the phosphate ester linkage at either the 3′- or 5′-OH group, accompanying the formation of an enzyme–phosphate–polynucleotide intermediate. The energy of the diester bond is conserved in the attachment, and the two ends of the nick are free to rotate relative to the helix axis in a reversible reaction. Hence, if the DNA has an excess or deficiency of turns, it is thermodynamically unstable relative to the nicked form. After untwisting, the diester bond is reformed at the nick as the untwisting enzyme becomes detached and regenerates the original closed-circular DNA molecule.

In the replication of small, single-stranded phage chromosomes, the enzymes used are "borrowed" from the host cell by the infecting virus. A general description (*247, 248*) of how some of these enzymes operate at the replicating fork is suggested in Figure 2.48, which includes other enzymes not mentioned here in detail.

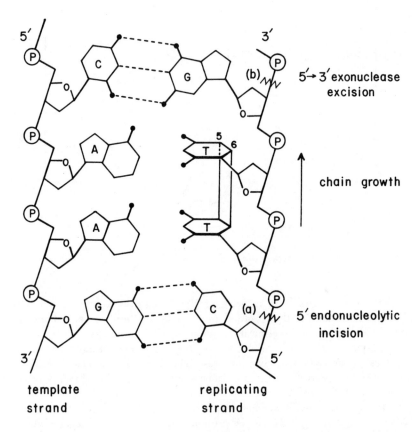

Figure 2.47. Expulsion of a thymine dimer from DNA by 5′ → 3′ exo-nuclease activity of Pol I. (a) Occurrence of a 5′ endonucleolytic incision near the lesion, thus allowing the dimer to peel away while DNA polymerase activity performs repair synthesis in the 5′ → 3′ direction. (b) Excision of the dimer by 5′ → 3′ nuclease activity of Pol I, followed by ligase activity to join the newly synthesized and original DNA chains.

All of these functions of DNA involve cleavage and the reformation of phosphoester linkages or both. Active-site geometries are largely unknown. However, mechanistic criteria available on nonenzymatic systems seem to be directly applicable in the mechanism proposed for chain elongation, based on the determination of the geometry of deoxynucleoside triphosphate substrates bound to *E. coli* DNA Pol I (230). It is expected that further structural work on some of the enzyme systems required for DNA synthesis will lead to related mechanisms in which nucleophilic attack at phosphate ester linkages proceeds in-line through postulated, near trigonal bipyramidal intermediates. Whether or not these mechanisms can be verified in vivo remains uncertain, especially since DNA functions established from in vitro studies

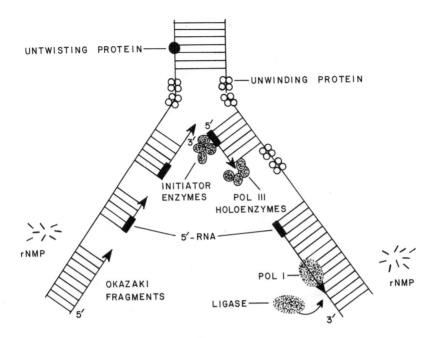

Figure 2.48. Some enzymes at the replicating fork in the template-directed DNA synthesis of small, single-stranded phage chromosomes. Pol III holo-enzyme catalyzes most of the DNA chain growth, initiated by the RNA primer fragment. Upon hydrolysis of the latter, the resulting gaps between Okazaki fragments (~1000 nucleotides long) are filled by Pol I. Ribonucleo-tide monophosphates (rNMP) are formed in the hydrolysis of the primer RNA. DNA ligase joins the base-paired fragments to form a continuous strand. Mg(II) and Zn(II) probably are required for polymerase activities. The scheme shown is a modification of that given in Ref. 219, p 221.

may not necessarily be the same as those occurring within the cell. For example, the role of cell membranes in DNA replication may alter DNA functions significantly.

Staphylococcal Nuclease. Like bovine pancreatic RNase A and RNase S (pp 191 and 202), staphylococcal nuclease hydrolyzes nucleic acids to produce 3′-nucleoside monophosphates *(249)* via a proposed trigonal bi-pyramidal transition state *(232, 239)*. In contrast, the RNases act only on RNA while staphylococcal nuclease operates on both RNA and DNA. This behavior is consistent with the lack of any evidence suggesting that the mechanism of hydrolysis of staphylococcal nuclease proceeds through cyclic phosphate intermediates like that supported for the first step of RNase action. For example, no hydrolysis of 2′, 3′-cyclic phosphate esters has been observed *(250)* with the staphylococcal enzyme.

Of the many known nucleases, staphylococcal nuclease is perhaps the the best characterized. It consists of a single polypeptide chain of 149 amino

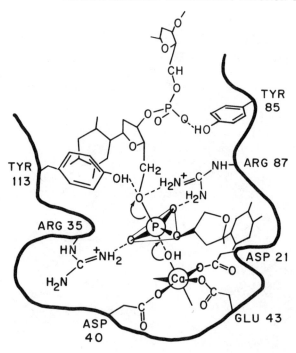

Figure 2.49. An in-line mechanism (239) suggested for staphylococcal nuclease (also see Ref 253)

acid residues with a molecular weight of 16,900 (*249*). The crystal structure determined earlier to about 2.5 Å resolution for the nulcease–thymidine 3′, 5′-diphosphate (pdTp)–calcium ion complex (*251, 252*) has been refined more recently to 1.5 Å (*253*). The crystal structure of the native enzyme is also available (*251*). Incisive NMR studies (*254, 255*) implicate two tyrosines at the active site by direct interaction with the latter nucleotide inhibitor, pdTp. Also, use of paramagnetic relaxation techniques (*256*) has characterized the active site for the related nuclease–pdTp–Gd(III) complex. The structural information (other than the more recent x-ray data), combined with chemical evidence (*249, 257, 258, 259*), has led to the mechanism (*239*) indicated in Figure 2.49.

The enzyme-bound Ca(II) ion, coordinated by three or four carboxylates, is 4.7 ±0.2 Å from the phosphorus atom in the nuclease–thymidine diphosphate (pdTp) complex. At this distance, an intervening water or hydroxyl ligand could be accommodated and lead to an in-line nucleophilic attack on phosphorus (*232, 239*). Both the presence of phosphate oxygens, strongly hydrogen-bonded to the two arginine–guanidinium cations (Arg 35 and Arg 87), and the presence of the coordinated metal ion promote an S_N2

mechanism by electron withdrawal at phosphorus and increase the nucleophilicity of an attacking water molecule (*232, 239*). The 1.5 Å data suggests (*253*) that Glu 43 has sufficient room for an intervening water molecule
between it and the Ca (II) ion. In this variant of the mechanism in Figure
2.49, an additional water molecule, bound between Glu 43 and an oxygen
atom of the 5′-phosphate, is promoted to an in-line nucleophilic attack by
enzyme constraints.

Interestingly, hydrolysis of deoxythymidine-3′-phosphate-5′-*p*-nitrophenyl-
phosphate (**CXV**) catalyzed by staphylococcal nuclease gave exclusive formation of *p*-nitrophenylphosphate, whereas nonenzymatic base hydrolysis of the
same substrate causes displacement of nitrophenoxide, a far better leaving
group than the 5′-oxyanion of deoxythymidine (*259*). These results suggest

Enzymatic hydrolysis of **CXV**   Nonenzymatic hydrolysis of **CXV**

(*259*) that catalysis by staphylococcal nuclease occurs within the framework
of established principles of pentacoordination. Consistent with the in-line
mechanism in Figure 2.49, the Thymidine Diphosphate **CXV** is most likely
located at the active site so that the better leaving group is not in a position
to depart nor can it rearrange to an axial position because of enzyme constraints.

In modeling staphylococcal nuclease action on the Substrate **CXV**, using
the molecular mechanics approach outlined for ribonuclease catalysis (*217*)
(p 199ff), we have found (*260*) (in agreement with the above), that the
pathway for enzymatic hydrolysis is of lower energy when the group being
displaced is the 5′-oxyanion of deoxythymidine. Furthermore, simulation of
the nonenzymatic hydrolysis supports the appearance of the better leaving
group, the nitrophenoxide ion, as the more favorable pathway rather than
deoxythymidine 3′-phosphate (*260*).

Of course, the inferences about the mechanism of action of staphylococcal
nuclease, like those that were made about enzymes discussed earlier in this

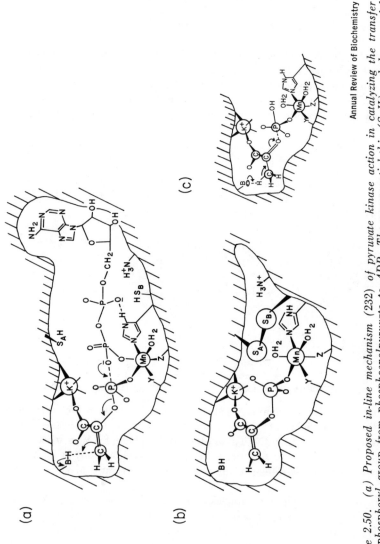

Annual Review of Biochemistry

Figure 2.50. (a) Proposed in-line mechanism (232) of pyruvate kinase action in catalyzing the transfer of a phosphoryl group from phosphoenolpyruvate to ADP. The nonessential thio (S_AH) and the essential thio (S_BH) are also indicated. (b) Proposed occlusion (232) of ADP in the disulfide modification of pyruvate kinase in which the thiols are linked. (c) Proposed in-line mechanism for the enolization of pyruvate (232).

chapter, assume a close correspondence between crystal and solution structures. Also, structural evidence obtained on bound inhibitor complexes is assumed to remain relatively unaltered when considering active-site geometries of hydrolyzable substrates. With regard to the former point, it is pleasing that the active-site geometry for the bound pdTp inhibitor deduced from paramagnetic relaxation measurements (*256*) using Gd(III) and that resulting from the x-ray study (*252*) of the corresponding Ca(II) complex gave general agreement to within the experimental uncertainty of the respective data (*256*), at least when the lower-resolution x-ray data were considered.

One further aspect of this well-studied nuclease system concerns studies that emphasize the importance of the intact enzyme conformation in maintaining full nuclease activity. In one such study (*249, 261*), a fragment consisting of the first 126 amino acid residues can form a complex through noncovalent interactions with a fragment containing residues 99–149. Some enzymatic activity is retained although only about 10% of that of the intact enzyme. Both of these fragments by themselves seem to be structureless in solution. Shortening by deleting residues 99–114 and 142–149 is possible without further apparent loss of activity- and structure-producing capacity.

Pyruvate Kinase. Structural work, primarily as-a result of NMR measurements (*232, 239, 262–269*) and chemical modification studies (*270, 271, 272*), led to a proposed mechanism (Figure 2.50) for pyruvate kinase, an enzyme that catalyzes the transfer of a phosphoryl group from phosphoenolpyruvate to adenosine-5′-diphosphate (ADP) (Figure 2.50a) as well as the reversible enolization of pyruvate (Figure 2.50c). This reaction represents the last step in glycolysis, i.e., the sequence of reactions that converts

Phosphoenolpyruvate   Pyruvate

glucose to pyruvate with the accompanying production of adenosine-5′-triphosphate (ATP) (*273*).

Based on ^{13}C and ^{31}P NMR relaxation measurements, both the carboxyl and carbonyl carbon atoms of pyruvate are determined to be 7.3 Å from the enzyme-bound Mn(II) ion while the lower limit of the Mn(II) to phosphorus distance is 4.5 Å (*266*). The mechanism in Figure 2.50 is consistent with these distances established for the active quaternary pyruvate kinase–Mn(II)–phosphate–pyruvate complex (*266*). Again, an in-line attack via trigonal bipyramidal intermediates is indicated. Hopefully, crystallographic data on cat muscle pyruvate kinase and on the enzyme–substrate complexes

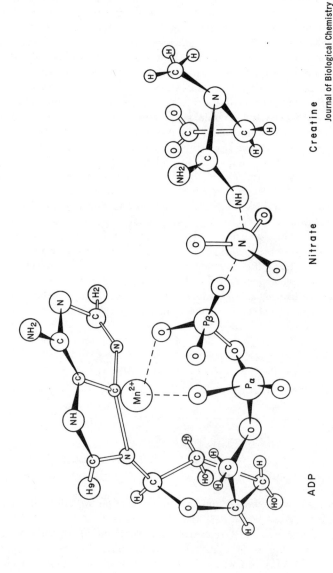

ADP   Nitrate   Creatine

Journal of Biological Chemistry

Figure 2.51.   Proposed structure (277) of the abortive quaternary complex of creatine kinase, Mn(II)—ADP—NO₃—creatine, which may serve as a transition-state analog (276) of the enzymatic reaction based on kinetic, EPR, and nuclear relaxation studies (239, 276, 277, 280)

with ADP and phosphoenol pyruvate (*274*) will provide additional insight into the mechanism.

Creatine Kinase. Analogous to pyruvate kinase, creatine kinase is an enzyme that catalyzes the transfer of a phosphoryl group to ADP to form ATP (*275*). In this case the transfer takes place from phosphocreatine. This compound has a higher group-transfer potential than ATP and serves as a reservoir of high-potential phosphoryl groups in muscle. The supply of phosphocreatine, however, is rapidly depleted in active muscle. As the ATP level drops, glycolysis, the citric acid cycle, and oxidative phosphorylation take over in contributing to the generation of ATP.

$$\begin{array}{c} \text{NH}_2 \\ | \\ {}^{2-}\text{O}_3\text{P}-\text{N}-\text{C}-\text{N}-\text{CH}_2\text{CO}_2{}^- + \text{ADP} + \text{H}^+ \xrightarrow{\quad \text{creatine} \quad \text{kinase} \quad} \\ | \quad + \quad | \\ \text{H} \qquad \text{CH}_3 \end{array}$$

Phosphocreatine

$$\begin{array}{c} \text{NH}_2 \\ | \\ \text{H}_2\text{N}-\text{C}-\text{N}-\text{CH}_2\text{CO}_2{}^- + \text{ATP} \\ + \quad | \\ \text{CH}_3 \end{array}$$

Creatine

Like pyruvate kinase, the active-site region of creatine kinase has been mapped using EPR and nuclear relaxation studies on enzyme–substrate complexes (*239, 276–281*). Consistent with available data, the structure of the abortive quaternary complex of creatine kinase (*276, 277, 280*), i.e., $\text{Mn(II)}-$ $\text{ADP}-\text{NO}_3{}^-$–creatine, is shown in Figure 2.51.

It is found that $\text{M(II)}-$ADP and creatine raise the affinity of the enzyme for nitrate (*276, 277, 278*) and, correspondingly, that nitrate raises the affinity of the enzyme for ADP and creatine (*276, 277, 279*). Hence, the quaternary complex may be a transition state analog (*276*) for the enzymatic reaction proceeding by an in-line S_N2 mechanism. Although, Mn(II) apparently does not bind to nitrate (the analog of the transferable phosphoryl group in the ground state), Mn(II) could migrate and bind in the transition state (*239*). In this respect the mechanism would be similar to that of pyruvate kinase (Figure 2.50).

Na⁺ and K⁺ Ion Transport Adenosine Triphosphatase, (Na⁺–K⁺ ATPase). The most widespread transport system in animal cells is the Na⁺–K⁺ pump that hydrolyzes ATP. The energy released in this tightly-coupled system is used to pump K⁺ ion in and Na⁺ ion out. Thus, a high concentration of K⁺ and a low concentration of Na⁺ exist within the cell relative to the external medium. This gradient is important in controlling the electrical

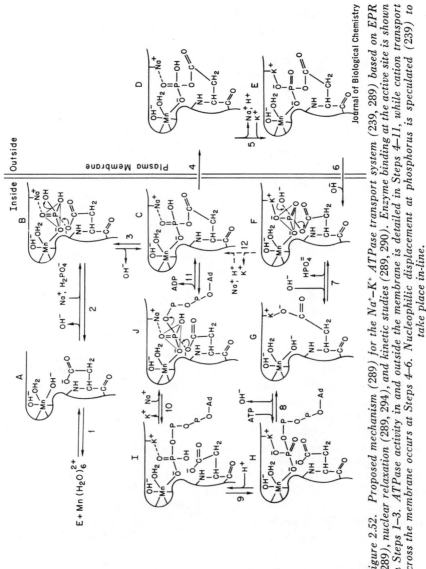

Figure 2.52. Proposed mechanism (289) for the Na⁺-K⁺ ATPase transport system (239, 289) based on EPR (289), nuclear relaxation (289, 294), and kinetic studies (289, 290). Enzyme binding at the active site is shown in Steps 1–3. ATPase activity in and outside the membrane is detailed in Steps 4–11, while cation transport across the membrane occurs at Steps 4–6. Nucleophilic displacement at phosphorus is speculated (239) to take place in-line.

characteristics of nerves and muscles and in the transport of sugars and amino acids across cell membranes (*282*).

The enzyme responsible for the active transport of the cations across the cell membrane is Na^+-K^+ ATPase. It is known (*283, 284*) that this enzyme can be phosphorylated in the presence of Mg^{2+} or Mn^{2+} by either ATP or P_i, the former requiring Na^+ ion (*285*) and the latter, K^+ ion (*286*). This suggests the two-step process:

$$E + ATP \xrightarrow{\ Na^+,\, M^{2+}\ } E - P + ADP$$

$$E - P + H_2O \xrightarrow{\ K^+,\, M^{2+}\ } E + P_i$$

The phosphoenzyme intermediate is an aspartyl phosphate (*287*) although phosphorylation of a glutamate side chain also has been reported (*288*). Cardiotonic steroids such as ouabain, when located outside the cell membrane, inhibit the dephosphorylation reaction. The latter action appears essential in the clinical use of digitalis, a cardiotonic steroid, in increasing the contraction of heart muscle.

Significant findings obtained by kinetic and magnetic resonance techniques establish that the enzyme binds Mn^{2+} at one tight binding site, identified as the active site, and that Na^+ is effective in causing the phosphate monoanion to coordinate to the enzyme-bound Mn^{2+} while K^+ is found to enhance ternary complex formation between the phosphate dianion and the enzyme-bound Mn^{2+} system (*289*). Therefore, conversion of a K^+-binding site to a Na^+-binding site may be achieved simply by protonation of the phosphoryl group bound at the active site of Na^+-K^+ ATPase.

The mechanism suggested in Figure 2.52 is consistent with all of the above results. Although not unique, it provides the simplest interpretation of the present data (*239, 283, 285, 286, 289–294*). For example, the formation of a binary Mn(II)–ATPase complex with four rapidly exchanging water protons on Mn(II) supports Step 1 (*289*). Proton relaxation-rate titrations support the formation of a complex with the phosphate monoanion (indicated in **B**) (*289*). Nucleophilic displacement of an OH^- group on phosphorus by the aspartyl carboxyl oxygen gives the Na^+-coordinated Intermediate **C**. In this form, Na^+ is transported through the plasma membrane. Subsequent loss of a proton converts the binding site from Na^+ to K^+ (**D** → **E**). The latter is consistent with the pH dependence of P_i binding, showing that Na^+ binds to the monoanion and K^+ to the dianion (*289*). These formulations agree with the order-of-magnitude weaker binding found for Na^+ compared with K^+ (*283*) since an alkali ion should be held more tightly by the dianion species (*289*). Formation of **E** yields the active-site complex by which K^+ is transported in the cell. Dephosphorylation via an in-line attack by OH^- ion leads to **G**, the enzyme–K^+ complex proposed by Post et al. (*290*). **H** is analogous to **E** except the phosphate dianion is replaced by ATP. Protonation of **H** converts the K^+ site to a Na^+ site, and **J** is formed. Analogous to **B**, attack by the aspartyl carboxyl group causes cleavage at the γ-phosphoryl group and completes the ATPase cycle. The latter phosphoryl group functions as a carrier in selectively transporting a Na^+ ion and a proton out of the cell and a K^+ ion into the cell in a mechanism driven by ATP hydrolysis.

Literature Cited

1. Westheimer, F. H., *Acc. Chem. Res.* (1968) **1**, 70.
2. Berry, R. S., *J. Chem. Phys.* (1960) **32**, 933.
3. Holmes, R. R., *Acc. Chem. Res.* (1979) **12**, 257.
4. Huheey, J. E., *J. Org. Chem.* (1966) **31**, 2365.
5. Swain, C. G., Lupton, E. C., Jr., *Am. Chem. Soc.* (1968) **90**, 4328.
6. Holmes, R. R., Deiters, J. A., *J. Am. Chem. Soc.* (1977) **99**, 3318.
7. Holmes, R. R., Deiters, J. A., *J. Chem. Res. (S)* (1977) 92.
8. Holmes, R. R., *J. Am. Chem. Soc.* (1974) **96**, 4143.
9. Bruice, T. C., Benkovic, S., "Biorganic Mechanisms," Vol. 2, pp 1–109, Benjamin, New York, 1966.
10. Kumamoto, J., Cox, J. R., Jr., Westheimer, F. H., *J. Am. Chem. Soc.* (1956) **78**, 4858.
11. Khorana, H. G., Tener, G. M., Wright, R. S., Moffatt, J. G., *J. Am. Chem. Soc.* (1957) **79**, 430.
12. Eberhard, A., Westheimer, F. H., *J. Am. Chem. Soc.* (1965) **87**, 253.
13. Brown, D. M., Todd, A. R., *J. Chem. Soc.* (1952) 52.
14. Cherbuliez, E., Probst, H., Rabinowitz, J., *Helv. Chim. Acta* (1959) **42**, 1377.
15. Wall, R. E., Ph.D. thesis, Harvard University, 1960.
16. Dennis, E. A., Westheimer, F. H., *J. Am. Chem. Soc.* (1966) **88**, 3432.
17. Aksnes, G., Bergesen, K., *Acta Chem. Scand.* (1966) **20**, 2508.
18. Covitz, F., Westheimer, F. H., *J. Am. Chem. Soc.* (1963) **85**, 1773.
19. Dennis, E. A., Westheimer, F. H., *J. Am. Chem. Soc.* (1966) **88**, 3431.
20. Newton, M. G., Cox, J. R., Jr., Bertrand, J. A., *J. Am. Chem. Soc.* (1966) **88**, 1503.
21. Kerst, F., Ph.D. thesis, Harvard University, 1967.
22. Ramirez, F., Madan, O. P., Desai, N. B., Meyerson, S., Banas, E. M., *J. Am. Chem. Soc.* (1963) **85**, 2681.
23. Haake, P. C., Westheimer, F. H., *J. Am. Chem. Soc.* (1961) **83**, 1102.
24. Kluger, R., Covitz, F., Dennis, E., Williams, L. D., Westheimer, F. H., *J. Am. Chem. Soc.* (1969) **91**, 6066.
25. Steitz, T. A., Lipscomb, W. N., *J. Am. Chem. Soc.* (1965) **87**, 2488.
26. Lide, D. R., Mann, D. E., *J. Chem. Phys.* (1958) **29**, 914.
27. Usher, D. A., Dennis, E. A., Westheimer, F. H., *J. Am. Chem. Soc.* (1965) **87**, 2320.
28. Gorenstein, D. G., Luxon, B. A., Findlay, J. B., Momii, R., *J. Am. Chem. Soc.* (1977) **99**, 4170.
29. Luckenbach, R., "Dynamic Stereochemistry of Pentacoordinated Phosphorus and Related Elements," p. 96ff, Georg Thieme Publishing Co., Stuttgart, 1973.
30. Gorenstein, D., *J. Am. Chem. Soc.* (1970) **92**, 644.
31. Green, M., Hudson, R. F., *J. Chem. Soc.* (1963) 540.
32. Boyd, D. B., *J. Am. Chem. Soc.* (1969) **91**, 1200.
33. Ramirez, F., Desai, N. B., *J. Am. Chem. Soc.* (1960) **82**, 2652.
34. Ramirez, F., *Acc. Chem. Res.* (1968) **1**, 168.
35. Ramirez, F., Madan, O. P., Heller, S. R., *J. Am. Chem. Soc.* (1965) **87**, 731.
36. Kukhtin, V. A., Kirillova, K. M., Shagidullin, R. R., Samitov, Yu. Yu., Lyazina, N. A., Rakova, N. F., *Zh. Obshch. Khim.* (1962) **32**, 2039.

37. Gorenstein, D., Westheimer, F. H., *J. Am. Chem. Soc.* (1967) **89**, 2762.
38. Hudson, R. F., Brown, C., *Acc. Chem. Res.* (1972) **5**, 204.
39. Gay, D. C., Hamer, N. K., *J. Chem. Soc., (London) Perkin Trans. II* (1972) 929.
40. Gay, D. C., Hamer, N. K., *Chem. Commun.* (1970) 1564.
41. Bel'skii, V. E., Bezzubova, N. N., Lustina, Z. V., Eliseenkov, V. N., Pudovik, A. N., *Zh. Obshch. Khim.* (1969) **39**, 181.
42. Van Wazer, J. R., "Phosphorus and Its Compounds," Vol. I, p. 887, Interscience, New York, 1958.
43. Huggins, M. L., *J. Am. Chem. Soc.* (1953) **75**, 4123, 4126.
44. Kluger, R., Westheimer, F. H., *J. Am. Chem. Soc.* (1969) **91**, 4143.
45. Kluger, R., Kerst, F., Lee, D. G., Dennis, E. A., Westheimer, F. H., *J. Am. Chem. Soc.* (1967) **89**, 3918.
46. Kluger, R., Kerst, F., Lee, D. G., Westheimer, F. H., *J. Am. Chem. Soc.* (1967) **89**, 3919.
47. Hawes, W., Trippett, S., *Chem. Commun.* (1968) 577.
48. *Ibid.* (1968) 295.
49. Aksnes, G., Bergesen, K., *Acta Chem. Scand.* (1965) **19**, 931.
50. Aksnes, G., Brudvik, L. J., *Acta Chem. Scand.* (1963) **17**, 1616.
51. McEwen, W. E., Kumli, K. F., Blade-Font, A., Zanger, M., VanderWerf, C. A., *J. Am. Chem. Soc.* (1964) **86**, 2378.
52. McEwen, W. E., Axelrad, G., Zanger, M., VanderWerf, C. A., *J. Am. Chem. Soc.* (1965) **87**, 3948.
53. Blade-Font, A., VanderWerf, C. A., McEwen, W. E., *J. Am. Chem. Soc.* (1960) **82**, 2396.
54. Kumli, K. F., McEwen, W. E., VanderWerf, C. A., *J. Am. Chem. Soc.* (1959) **81**, 3805.
55. Zanger, M., VanderWerf, C. A., McEwen, W. E., *J. Am. Chem. Soc.* (1959) **81**, 3806.
56. Cremer, S. E., Trivedi, B. C., Weitl, F. L., *J. Org. Chem.* (1971) **36**, 3226.
57. Frank, D. S., Usher, D. A., *J. Am. Chem. Soc.* (1967) **89**, 6360.
58. Ramirez, F., Hansen, B., Desai, N. B., *J. Am. Chem. Soc.* (1962) **84**, 4588.
59. Archer, A. A. P. G., Harley-Mason, J., *Proc. Chem. Soc.* (1958) 285.
60. Lieske, C. N., Miller, E. G., Jr., Zeger, J. J., Steinberg, G. M., *J. Am. Chem. Soc.* (1966) **88**, 188.
61. Cox, J. R., Jr., Ramsay, O. B., *Chem. Rev.* (1964) **64**, 317.
62. Witzel, H., Botta, A., Dimroth, K., *Chem. Ber.* (1965) **98**, 1465.
63. De'Ath, N. J., Trippett, S., *Chem. Commun.* (1969) 172.
64. Lewis, R. A., Naumann, K., DeBruin, K. E., Mislow, K., *Chem. Commun.* (1969) 1010.
65. Corfield, J. R., De'Ath, N. J., Trippett, S., *J. Chem. Soc. C* (1971) 1930.
66. Luckenbach, R., *Phosphorus* (1972) **1**, 223.
67. *Ibid.* p. 229.
68. *Ibid.* p. 293.
69. Op. cit. in Ref. *29*, pp 109, 117.
70. Allen, D. W., Hutley, B. G., Mellor, M. T. J., *J. Chem. Soc., Perkin II* (1972) 63.
71. Parisek, C. P., McEwen, W. E., VanderWerf, C. A., *J. Am. Chem. Soc.* (1960) **82**, 5503.
72. De'Ath, N. J., Ellis, K., Smith, D. J. H., Trippett, S., *Chem. Commun.* (1971) 714.
73. DeBruin, K. E., Johnson, D. M., *J. Am. Chem. Soc.* (1973) **95**, 4675.
74. Farnham, W. B., Mislow, K., Mandel, N., Donahue, J., *J. Chem. Soc., Chem. Commun.* (1972) 120.
75. DeBruin, K. E., Johnson, D. M., *J. Chem. Soc., Chem. Commun.* (1975) 753.
76. DeBruin, K. E., Johnson, D. M., *Phosphorus*, (1974) **4**, 13.

77. Reiff, L. P., Szafraniec, L. J., Aaron, H. S., *Chem. Commun.* (1971) 366.
78. DeBruin, K. E., Johnson, D. M., *Phosphorus* (1974) **4**, 17.
79. Hine, J., Hine, M., *J. Am. Chem. Soc.* (1952) **74**, 5266.
80. Bergesen, K., *Acta Chem. Scand.* (1967) **21**, 1587.
81. Cremer, S. E., Trivedi, B. C., *J. Am. Chem. Soc.* (1969) **91**, 7200.
82. Hawes, W., Trippett, S., *J. Chem. Soc. C* (1969) 1465.
83. -Ul-Haque, M., Caughlan, C. N., *Chem. Commun.* (1968) 1228.
84. Op. cit. in Ref. *81*, Footnote 9.
85. Op. cit. in Ref. *81*, Footnote 3.
86. Fishwick, S. E., Flint, J., Hawes, W., Trippett, S., *Chem. Commun.* (1967) 1113.
87. Op. cit. in Ref. *29*, p. 143.
88. Ezzell B. R., *J. Org. Chem.* (1970) **35**, 2426.
89. Cremer, S. E., Chorvat, R. J., Trividi, B. C., *Chem. Commun.* (1969) 769.
90. Corfield, J. R., Shutt, J. R., Trippett, S., *Chem. Commun.* (1969) 789.
91. Schneider, W. P., *Chem. Commun.* (1969) 785.
92. Chioccola, G., Daly, J. J., *J. Chem. Soc. A* (1968) 568.
93. McEwen, W. E., Blade-Font A., VanderWerf, C. A., *J. Am. Chem. Soc.* (1962) **84**, 677.
94. McEwen, W. E., Wolf, A. P., VanderWerf, C. A., Blade-Font, A., Wolfe, J. W., *J. Am. Chem. Soc.* (1967) **89**, 6685.
95. DeBruin, K. E., Naumann, K., Zon, G., Mislow, K., *J. Am. Chem. Soc.* (1969) **91**, 7031.
96. Naan, M. P., Powell, R. L., Hall, C. D., *J. Chem. Soc. B* (1971) 1683.
97. Grayson, M., Farley, C. E., *Chem. Commun.* (1967) 831.
98. Omelanczuk, J., Mikolajczyk, M., *Tetrahedron* (1971) **27**, 5587.
99. DeBruin, K. E., Padilla, A. G., Campbell, M.-T., *J. Am. Chem. Soc.* (1973) **95**, 4681.
100. Hoffman, R., Howell, J. M., Muetterties, E. L., *J. Am. Chem. Soc.* (1972) **94**, 3047.
101. Bone, S., Trippett, S., Whittle, P. J., *J. Chem. Soc., Perkin I* (1974) 2125.
102. Oram, R. K., Trippett, S., *J. Chem. Soc., Perkin I* (1973) 1300.
103. Howard, J. A., Russell, D. R., Trippett, S., *J. Chem. Soc., Chem. Commun.* (1973) 856.
104. Marsi, K. L., *J. Org. Chem.* (1975) **40**, 1779.
105. Marsi, K. L., *J. Am. Chem. Soc.* (1969) **91**, 4724.
106. Egan, W., Chauviere, G., Mislow, K., Clark, R. T., Marsi, K. L., *Chem. Commun.* (1970) 733.
107. Marsi, K. L., Burns, F. B., Clark, R. T., *J. Org. Chem.* (1972) **37**, 238.
108. Fenton, G. W., Ingold, C. K., *J. Chem. Soc.* (1929) 2342.
109. Hey, L., Ingold, C. K., *J. Chem. Soc.* (1933) 531.
110. Cavell, R. G., Poulin, D. D., The, K. I., Tomlinson, A. J., *J. Chem. Soc., Chem. Commun.* (1974) 19.
111. Cavell, R. G., Gibson, J. A., The, K. I., *J. Am. Chem. Soc.* (1977) **99**, 7841.
112. DeBruin, K. E., Zon, G., Naumann, K., Mislow, K., *J. Am. Chem. Soc.* (1969) **91**, 7027.
113. Cremer, S. E., *Chem. Commun.* (1970) 616.
114. Corfield, J. R., De'Ath, N. J., Trippett, S., *Chem. Commun.* (1970) 1502.
115. Horner, L., *Pure Appl. Chem.* (1964) **9**, 225.
116. Naumann, K., Zon, G., Mislow, K., *J. Am. Chem. Soc.* (1969) **91**, 7012.
117. Mislow, K., *Acc. Chem. Res.* (1970) **3**, 321.
118. Zon, G., Mislow, K., *Fortschr. Chem. Forsch.* (1971) **19**, 61.
119. Marsi, K. L., *Chem. Commun.* (1968) 846.
120. Marsi, K. L., Oberlander, J. E., *J. Am. Chem. Soc.* (1973) **95**, 200.
121. Marsi, K. L., Clark, R. T., *J. Am. Chem. Soc.* (1970) **92**, 3791.

122. Marsi, K. L., *J. Am. Chem. Soc.* (1971) **93**, 6341.
123. Op. cit. in Ref. *29*, pp 158-9.
124. Fitzgerald, A., Smith, G. D., Caughlan, C. N., Marsi, K. L., Burns, F. B., *J. Org. Chem.* (1976) **41**, 1155.
125. Op. cit. in Ref. *29*, p 164.
126. Holmes, R. R., *J. Am. Chem. Soc.* (1975) **97**, 5379.
127. Allen, D. W., Millar, I. T., *Chem. Ind.* (1967) 2178.
128. Allen, D. W., Millar, I. T., Mann, F. G., *J. Chem. Soc. C* (1967) 1869.
129. Allen, D. W., Millar, I. T., *J. Chem. Soc. B* (1969) 263.
130. Allen, D. W., Tebby, J. C., *J. Chem. Soc. B* (1970) 1527.
131. Allen, D. W., Millar, I. T., *J. Chem. Soc. C* (1969) 252.
132. Albrand, J.-P., Gagnaire, D., Picard, M., Robert, J.-B. *Tetrahedron Lett.* (1970) 4593.
133. Priestley, H. J., Snyder, J. P., *Tetrahedron Lett.* (1971) 2433.
134. Bergesen, K., *Acta Chem. Scand.* (1966) **20**, 899.
135. Op. cit. in Ref. *29*, pp 173–179 and references cited therein.
136. Brophy, J. J., Gallagher, M. J., *Chem. Commun.* (1967) 344.
137. Driver, G. E., Gallagher, M. J., *Chem. Commun.* (1970) 150.
138. Cuddy, B. D., Murray, J. C. F., Walker, B. J., *Tetrahedron Lett.* (1971) 2397.
139. Martin, J. C., Perozzi, E. F., *Science* (1976) **191**, 154.
140. Martin, L. D., Perozzi, E. F., Martin, J. C., *J. Am. Chem. Soc.* (1979) **101**, 3595.
141. Smith, D. J. H., Trippett, S., *Chem. Commun.* (1969) 855.
142. Cremer, S. E., Chorvat, R. J., *J. Org. Chem.* (1967) **32**, 4066.
143. Op. cit. in Ref. *29*, p 150 for additional examples.
144. Kyba, E. P., *J. Am. Chem. Soc.* (1975) **97**, 2554.
145. Kyba, E. P., *J. Am. Chem. Soc.* (1976) **98**, 4805.
146. Kyba, E. P., Hudson, C. W., *Tetrahedron Lett.* (1975) 1869.
147. Greenhalgh, R., Hudson, R. F., *Chem. Commun.* (1968) 1300.
148. Greenhalgh, R., Hudson, R. F., *Phosphorus* (1972) **2**, 1.
149. Aksnes, G., Eriksen, R., *Acta Chem. Scand.* (1966) **20**, 2463.
150. Panar, M., Kaiser, E. T., Westheimer, F. H., *J. Am. Chem. Soc.* (1963) **85**, 602.
151. Aksnes, G., Aksnes, D., *Acta Chem. Scand.* (1964) **18**, 38.
152. Evdakov, V. P., Mizrakh, L. I., Sandalova, L. Y., *Dokl. Akad. Nauk. SSSR* (1965) **162**, 573.
153. Evdakov, V. P., Mizrakh, L. I., Sandalova, L. Y., *Zh. Obshch. Khim.* (1965) **35**, 1314.
154. Burgada, R., Houalla, D., Wolf, R., *C. R. Acad. Sci., Ser. C* (1967) 356.
155. Hudson, R. F., Searle, R. J. G., *J. Chem. Soc. B* (1968) 1349.
156. Hudson, R. F., Mancuso, A., *Chem. Commun.* (1969) 522.
157. Campbell, T. W., Monagle, J. J., Foldi, V. S., *J. Am. Chem. Soc.* (1962) **84**, 3673.
158. Monagle, J. J., Campbell, T. W., McShane, H. F., *J. Am. Chem. Soc.* (1962) **84**, 4288.
159. Monagle, J. J., *J. Org. Chem.* (1962) **27**, 3851.
160. Bentrude, W. G., "Free Radicals," Vol. 2 Ch. 22, J. K. Kochi, Ed., Wiley-Interscience, New York, 1973.
161. Davies, A. G., Roberts, B. P., "Free Radicals," Vol. I, Ch. 10, J. K. Kochi, Ed., Wiley-Interscience, New York, 1973.
162. Bentrude, W. G., Hargis, J. H., Rusek, P. E., Jr., *J. Chem. Soc., Chem. Commun.* (1969) 296.
163. Bentrude, W. G., Wielesek, R. A., *J. Am. Chem. Soc.* (1969) **91**, 2406.
164. Buckler, S. A., *J. Am. Chem. Soc.* (1962) **84**, 3093.

165. Bentrude, W. G., Hansen, E. R., Khan, W. A., Rogers, P. E., *J. Am. Chem. Soc.* (1972) **94**, 2867.
166. Davies, A. G., Griller, D., Roberts, B. P., *J. Chem. Soc., Perkin II* (1972) 2224.
167. Watts, G. B., Griller, D., Ingold, K. U., *J. Am. Chem. Soc.* (1972) **94**, 8784.
168. Bentrude, W. G., Rogers, P. E., *J. Am. Chem. Soc.* (1976) **98**, 1674.
169. Bentrude, W. G., Hansen, E. R., Khan, W. A., Min, T. B., Rogers, P. E., *J. Am. Chem. Soc.* (1973) **95**, 2286.
170. Boekestein, G., Jansen, E. H. J. M., Buck, H. M., *J. Chem. Soc., Chem. Commun.* (1974) 118.
171. Davies, A. G., Griller, D., Roberts, B. P., *J. Chem. Soc., Perkin II* (1972) 993.
172. Op. cit. in Ref. *42*, p 33.
173. Davies, A. G., Dennis, R. W., Griller, D., Roberts, B. P., *J. Organomet. Chem.* (1972) **40**, C33.
174. Krusic, P. J., Mahler, W., Kochi, J. K., *J. Am. Chem. Soc.* (1972) **94**, 6033.
175. Davies, A. G., Dennis, R. W., Roberts, B. P., *J. Chem. Soc., Perkin II* (1974) 1101.
176. Bentrude, W. G., Hargis, J. H., Johnson, N. A., Min, T. B., Rusek, P. E., Jr., Tan, H.-W., Wielesek, R. A., *J. Am. Chem. Soc.* (1976) **98**, 5348.
177. Dennis R. W., Roberts, B. P., *J. Chem. Soc., Perkin II* (1975) 140.
178. Bentrude, W. G., Min, T. B., *J. Am. Chem. Soc.* (1976) **98**, 2918.
179. Bentrude, W. G., Alley, W. D., Johnson, N. A., Murakami, M., Nishikida, K., Tan, H.-W., *J. Am. Chem. Soc.* (1977) **99**, 4383.
180. Nakanishi, A., Nishikida, K., Bentrude, W. G., *J. Am. Chem. Soc.* (1978) **100**, 6403, and references cited therein.
181. Bentrude, W. G., Fu, J-J. L., Rogers, P. E., *J. Am. Chem. Soc.* (1973) **95**, 3625.
182. Horner, L., Haufe, J., *Chem. Ber.* (1968) **101**, 2903.
183. Rothius, R., Luderer, T. K. J., Buck, H. M., *Rec. Trav. Chim. Pays-Bas.* (1972) **91**, 836.
184. Pauling, L., *Nature* (1948) **161**, 707.
185. Richards, F. M., Wyckoff, H. R., "The Enzymes," P. D. Boyer, Ed., 3rd ed., Vol. IV, pp 647–806, Academic Press, New York, 1971.
186. Dickerson, R. E., Geis, I., "The Structure and Action of Proteins," pp 79–81, Harper Row, New York, 1969.
187. Avey, H. P., Boles, M. O., Carlisle, C. H., Evans, S. A., Morris, S. J., Palmer, R. A., Woolhouse, B. A., Shall, S. *Nature* (1967) **213**, 557.
188. Kartha, G., Bello, J., Harker, D., *Nature* (1967) **213**, 862.
189. Carlisle, C. H., Palmer, R. A., Mazumdar, S. K., Gorinsky, B. A., Yeates, D. G. R., *J. Mol. Biol.* (1974) **85**, 1.
190. Wyckoff, H. W., Hardman, K. D., Allewell, N. M., Inagami, T., Tsernoglou, D., Johnson, L. N., Richards, F. M., *J. Biol. Chem.* (1967) **242**, 3749.
191. Wyckoff, H. W., Hardman, K. D., Allewell, N. M., Inagami, T., Johnson, L. N., Richards, F. M., *J. Biol. Chem.* (1967) **242**, 3984.
192. Wyckoff, H. W., Tsernoglou, D., Hanson, A. W., Knox, J. R., Lee, B., Richards F. M., *J. Biol. Chem.* (1970) **245**, 305.
193. Richards, F. M., Wyckoff, H. W., "Atlas of Molecular Structures in Biology. 1. Ribonuclease –S," D. C. Phillips and F. M. Richards, Eds., Clarendon, Oxford, 1973.
194. Roberts, G. C. K., Dennis, E. A., Meadows, D. H., Cohen, J. S., Jardetzky, O., *Proc. Natl. Acad. Sci. U.S.A.* (1969) **62**, 1151.
195. Usher, D. A., *Proc. Natl. Acad. Sci. U.S.A.* (1969) **62**, 661.
196. Hummel, J. P., Kalnitsky, G., *Annu. Rev. Biochem.* (1964) **33**, 15.
197. Scheraga, H. A., *Federation Proc.* (1967) **26**, 1380.
198. Usher, D. A., Richardson, D. I., Jr., Eckstein, F., *Nature* (1970) **228**, 663.
199. Saenger, W., Eckstein, F., *J. Am. Chem. Soc.* (1970) **92**, 4712.

200. Eckstein, F., *J. Am. Chem. Soc.* (1970) **92**, 4718.
201. Brown, D. M., Dekker, C. A., Todd, A. R., *J. Chem. Soc.* (1952) 2715.
202. Usher, D. A., Erenrich, E. S., Eckstein, F., *Proc. Natl. Acad. Sci. U.S.A.* (1972) **69**, 115.
203. Harris, M. R., Usher, D. A., Albrecht, H. P., Jones, G. H., Moffatt, J. G., *Proc. Natl. Acad. Sci. U.S.A.* (1969) **63**, 246.
204. Abrash, H. I., Cheung, C. S., Davis, J. C., *Biochemistry* (1967) **6**, 1298.
205. Eckstein, F., *F.E.B.S. Lett.* (1968) **2**, 85.
206. Eckstein, F., Gindl, H., *Chem. Ber.* (1968) **101**, 1670.
207. Holmes, R. R., *Int. J. Peptide Protein Res.* (1976) **8**, 445.
208. Findlay, D., Herries, D. G., Mathias, A. P., Rabin, B. R., Ross, C. A., *Biochem. J.* (1962) **85**, 152.
209. Op. cit. in Ref. *185*, pp 775–776.
210. Follmann, H., Wieker, H.-J., Witzel, H., *Eur. J. Biochem.* (1967) **1**, 243.
211. Stryer, L., "Biochemistry," pp 124–129, W. H. Freeman and Co., San Francisco, 1975.
212. Gassen, H. G., Witzel, H., *Eur. J. Biochem.* (1967) **1**, 36.
213. Richards, F. M., Wyckoff, H. W., Allewell, N., "The Neurosciences: Second Study Program," p 970, F. O. Schmitt, Ed., Rockefeller University 1970.
214. Pauling, L., "The Nature of the Chemical Bond," 3rd ed., p 477, Cornell University Press, Ithaca, N.Y., 1960.
215. Sussman, J. L., Seeman, N. C., Kim, S. H., Berman, H., *J. Mol. Biol.* (1972) **66**, 403.
216. Griffin, J. H., Schechtes, A. N., Cohen, J. S., *Ann. N.Y. Acad. Sci.* (1973) **222**, 693.
217. Holmes, R. R., Deiters, J. A., Gallucci, J. C., *J. Am. Chem. Soc.* (1978) **100**, 7393.
218. Coulter, C. L., *J. Am. Chem. Soc.* (1973) **95**, 570.
219. Kornberg, A., "DNA Synthesis," W. H. Freeman, San Francisco, 1974.
220. Gefter, M. L., *Annu. Rev. Biochem.* (1975) **44**, 45.
221. Stryer, L., "Biochemistry," W. H. Freeman, San Francisco, 1975.
222. Mildvan, A. S., *Acc. Chem. Res.* (1977) **10**, 246.
223. Smith, D. W., *Progr. Biophys. Mol. Biol.* (1973) **26**, 321.
224. Setlow, R. B., Setlow, J. K., *Annu. Rev. Biophys. Bioeng.* (1972) **1**, 293.
225. Chamberlin, M. J., *Annu. Rev. Biochem.* (1974) **43**, 721.
226. Sobell, H. M., *Progr. Nucl. Acid Res. Mol. Biol.* (1973) **13**, 153.
227. Op. cit. in Ref. *221*, Chapter 23.
228. Sugino, A., Okazaki, R., *J. Mol. Biol.* (1972) **64**, 61.
229. Sugino, A., Okazaki, R., *Proc. Natl. Acad. Sci. U.S.A.* (1973) **70**, 88.
230. Sloan, D. L., Loeb, L. A., Mildvan, A. S., Feldmann, R. J., *J. Biol. Chem.* (1975) **250**, 8913.
231. Sundaralingam, M., *Biopolymers* (1969) **7**, 821.
232. Mildvan, A. S., *Annu. Rev. Biochem.* (1974) **43**, 357.
233. Trautner, T. A., Swartz, M. N., Kornberg, A., *Proc. Natl. Acad. Sci. U.S.A.* (1962) **48**, 449.
234. Op. cit. in Ref. *219*, p 79.
235. Abbould, M. M., Sim, W. J., Loeb, L. A., Mildvan, A. S., *J. Biol. Chem.* (1978) **253**, 3415.
236. Slater, J. P., Tamir, I., Loeb, L. A., Mildvan, A. S., *J. Biol. Chem.* (1972) **247**, 6784.
237. Slater, J. P., Mildvan, A. S., Loeb, L. A., *Biochem. Biophys. Res. Commun.* (1971) **44**, 37.
238. Springgate, C. F., Mildvan, A. S., Abramson, R., Engle, J. L., Loeb, L. A., *J. Biol. Chem.* (1973) **248**, 5987.
239. Mildvan, A. S., Grisham, C. M., *Struct. Bonding* (1974) **20**, 1.

240. Op. cit. in Ref. *221*, p 602.
241. Op. cit. in Ref. *219*, pp 85, 94.
242. Op. cit. in Ref. *220*, p 50.
243. Op. cit. in Ref. *221*, Chapter 23.
244. Op. cit. in Ref. *219*, p 203.
245. Wang, J. C., *J. Mol. Biol.* (1971) **55**, 523.
246. Champoux, J. J., Dulbecco, R., *Proc. Natl. Acad. Sci. U.S.A.* (1972) **69**, 143.
247. Op. cit. in Ref. *219*, p 221.
248. Op. cit. in Ref. *220*, p 51.
249. Anfinsen, C. B., Cuatrecasas, P., Taniuchi, H., 'The Enzymes," P. D. Boyer, Ed., 3rd ed., Vol. IV pp 177–204, Academic Press, New York, 1971.
250. Alexander, M., Heppel, L. A., Hurwitz, J., *J. Biol. Chem.* (1961) **236**, 3014.
251. Cotton, F. A., Hazen, E. E., Jr., 'The Enzymes," P. D. Boyer, Ed., 3rd ed., Vol. IV pp 153–176, Academic Press, New York, 1971.
252. Arnone, A., Bier, C. J., Cotton, F. A., Day, V. W., Hazen, E. E., Jr., Richardson, D. C., Richardson, J. S., Yonath, A., *J. Biol. Chem.* (1971) **246**, 2302.
253. Cotton, F. A., Hazen, E. E., Jr., Legg, M. J., *Proc. Natl. Acad. Sci. U.S.A.* (1979) **76**, 2551 and references cited therein.
254. Markley, J. L., Jardetzky, O., *J. Mol. Biol.* (1970) **50**, 223.
255. Roberts, G. C. K., Jardetzky, O., *Adv. Prot. Chem.* (1970) **24**, 447.
256. Furie, B., Griffen, J. H., Feldmann, R. J., Sokoloski, A., Schechter, A. N., *Proc. Natl. Acad. Sci. U.S.A.* (1974) **71**, 2833.
257. Chaiken, I. M., Anfinsen, C. B., *J. Biol. Chem.* (1971) **246**, 2285.
258. Sanchez, G. R., Chaiken, I. M., Anfinsen, C. B., *J. Biol. Chem.* (1973) **248**, 3653.
259. Dunn, B. M., DiBello, C., Anfinsen, C. B., *J. Biol. Chem.* (1973) **248**, 4769.
260. Deiters, J. A., Holmes, R. R., 178th Natl. Meet. of the Am. Chem. Soc., Washington, D.C., September, 1979, Abstracts No. BIOL 33.
261. Taniuchi, H., Anfinsen, C. B., *J. Biol. Chem.* (1971) **246**, 2291.
262. Mildvan, A. S., Nowak, T., Fung, C. H., *Ann. N.Y. Acad. Sci.* (1973) **222**, 192.
263. Nowak, T., Mildvan, A. S., *Biochemistry* (1972) **11**, 2819.
264. Mildvan, A. S., Leigh, J. S., Cohn, M., *Biochemistry* (1967) **6**, 1805.
265. Mildvan, A. S., Cohn, M., *J. Biol. Chem.* (1966) **241**, 1178.
266. Fung, C. H., Mildvan, A. S., Allerhand, A., Komoroski, R., Scrutton, M. C., *Biochemistry* (1973) **12**, 620.
267. Reuben, J., Kayne, F., *J. Biol. Chem.* (1971) **246**, 6227.
268. Kayne, F. J., Reuben, J., *J. Am. Chem. Soc.* (1970) **92**, 220.
269. Nowak, T., *J. Biol. Chem.* (1973) **248**, 7191.
270. Hollenberg, P. F., Flashner, M., Coon, M. J., *J. Biol. Chem.* (1971) **246**, 946.
271. Flashner, M., Hollenberg, P. F., Coon, M. J., *J. Biol. Chem.* (1972) **247**, 8114.
272. Flashner, M., Tamir, I., Mildvan, A. S., Meloche, H. P., Coon, M. J., *J. Biol. Chem.* (1973) **248**, 3419.
273. Op. cit. in Ref. *221*, Chapter 12.
274. Muirhead, H., Stammers, D. K., *Biochem. Soc. Trans.* (1974) **2**, 49.
275. Op. cit. in Ref. *221*, p 836.
276. Milner-White, E. J., Watts, D. C., *Biochem. J.* (1971) **122**, 727.
277. Reed, G., Cohn, M., *J. Biol. Chem.* (1972) **247**, 3073.
278. Reed, G. H., McLaughlin, A. C., *Ann. N.Y. Acad. Sci.* (1973) **222**, 118.
279. McLaughlin, A. C., Cohn, M., Kenyon, G. L., *J. Biol. Chem.* (1972) **247**, 4382.
280. Leigh, J. S., Jr., Ph.D. thesis, University of Pennsylvania, 1971.
281. Cohn, M., Leigh, J. S., Jr., Reed, G. H., *Cold Spring Harbor Symp. Quant. Biol.* (1971) **36**, 533.
282. Op. cit. in Ref. *221*, p 769.

283. Skou, J. C., *Biochim. Biophys. Acta* (1960) **42,** 6.
284. Atkinson, A., Hunt, S., Lowe, A. G., *Biochim. Biophys. Acta* (1968) **167,** 469.
285. Post, R. L., Sen, A. K., Rosenthal, A. S., *J. Biol. Chem.* (1965) **240,** 1437.
286. Lindenmayer, G. E., Laughter, A. H., Schwartz, A., *Arch. Biochem. Biophys.* (1968) **127,** 187.
287. Post, R. L., Kume, S. *J. Biol. Chem.* (1973) **248,** 6993.
288. Kahlenberg, A., Galsworthy, P. R., Hokin, L. E., *Arch. Biochem. Biophys.* (1968) **126,** 331.
289. Grisham, C. M., Mildvan, A. S., *J. Biol. Chem.* (1974) **249,** 3187.
290. Post, R. L., Hegyvary, C., Kume, S., *J. Biol. Chem.* (1972) **247,** 6530.
291. Post, R. L., Kume, S., Tobin, T., Orcutt, B., Sen, A. K., *J. Gen. Physiol.* (1969) **54,** 306s.
292. Neufeld, A. H., Levy, H. M., *J. Biol. Chem.* (1970) **245,** 4962.
293. Dahms, A. S., Kanazawa, T., Boyer, P. D., *J. Biol. Chem.* (1973) **248,** 6592.
294. Grisham, C. M., Mildvan, A. S., *Federation Proc.* (1974) **33,** 1331.

Index

Index